Mechanics and Physics of Precise Vacuum Mechanisms

FLUID MECHANICS AND ITS APPLICATIONS
Volume 91

Series Editor: R. MOREAU
MADYLAM
Ecole Nationale Supérieure d'Hydraulique de Grenoble
Boîte Postale 95
38402 Saint Martin d'Hères Cedex, France

Aims and Scope of the Series

The purpose of this series is to focus on subjects in which fluid mechanics plays a fundamental role.

As well as the more traditional applications of aeronautics, hydraulics, heat and mass transfer etc., books will be published dealing with topics which are currently in a state of rapid development, such as turbulence, suspensions and multiphase fluids, super and hypersonic flows and numerical modeling techniques.

It is a widely held view that it is the interdisciplinary subjects that will receive intense scientific attention, bringing them to the forefront of technological advancement. Fluids have the ability to transport matter and its properties as well as to transmit force, therefore fluid mechanics is a subject that is particularly open to cross fertilization with other sciences and disciplines of engineering. The subject of fluid mechanics will be highly relevant in domains such as chemical, metallurgical, biological and ecological engineering. This series is particularly open to such new multidisciplinary domains.

The median level of presentation is the first year graduate student. Some texts are monographs defining the current state of a field; others are accessible to final year undergraduates; but essentially the emphasis is on readability and clarity.

For other titles published in this series, go to
www.springer.com/series/5980

E.A. Deulin • V.P. Mikhailov • Yu.V. Panfilov
R.A. Nevshupa

Mechanics and Physics of Precise Vacuum Mechanisms

Springer

E.A. Deulin
Bauman Moscow State Technical University
2nd Baumanskaya 5
Moskva 107005
Russia
eadeulin@rdm.ru

V.P. Mikhailov
Bauman Moscow State Technical University
2nd Baumanskaya 5
Moskva 107005
Russia

Yu.V. Panfilov
Bauman Moscow State Technical University
2nd Baumanskaya 5
Moskva 107005
Russia

R.A. Nevshupa
Bauman Moscow State Technical University
2nd Baumanskaya 5
Moskva 107005
Russia

This book is a revised and updated translation of the original Russian work: Mechanics and Physics of Precise Vacuum Mechanisms (2 vols.), Moscow, NPK "Intelvac", 2000, 2001.

ISSN 0926-5112
ISBN 978-90-481-2519-7 e-ISBN 978-90-481-2520-3
DOI 10.1007/978-90-481-2520-3
Springer Dordrecht Heidelberg London New York

Library of Congress Control Number: 2009939688

Printed on acid-free paper

Springer is part of Springer Science+Business Media (www.springer.com)

Contents

Preface

The idea for this book was born between 1980–1985 when electronics, vacuum technology and mechanical engineering in the former USSR developed into worthy competitors in the global engineering industry and in the international scientific world. After the transition took place in Russia in the 1990s and to the moment of Russia's new industry establishing, many scientific results, many talented engineers were lost to the world science.

The authors of this book collected and accumulated the most interesting results in the field of engineering and science of the former USSR. Up till now these results are still at the highest scientific and engineering level in the world. These results can be interesting to scientists and engineers in the field of vacuum science and engineering, microelectronics technology, nanotechnologies, fusion experimental physics and other.

The main topics presented in this work are:

- the regulation and determination method of micro-particle flows generated by vacuum mechanisms in vacuum equipment and in electronics;
- the precise mechanisms of nanoscale precision based on magnetic and electric rheology phenomena;
- precise harmonic rotating UHV feedthroughs and non-coaxial nut-screw linear-motion vacuum feedthroughs, at unique technical characteristics;
- elastically deformed vacuum multi-coordinate motion feedthroughs without friction couples in vacuum;
- the computer system for failure prediction of vacuum mechanisms.

Chapters 1 and 2 show the typical vacuum equipment. They also show different types of precise vacuum drives. Chapter 3 considers the physical nature of different phenomena occurring during surface friction in vacuum. Chapter 4 shows a matrix system for the classification of vacuum drives and considers the system for estimation of driver parameters. Furthermore, it deals with the new principle of drive invention of a new type, which has to meet a number of complex technical requirements.

The original matrix method of mechanism analysis presented here allows us to determine the potential properties of the mechanisms on the design stage, to choose the mechanisms with required properties and to generate new mechanisms with the best properties. Chapter 5 considers the theory of vacuum mechanism precision. It also shows the analysis of the ways in which precision of vacuum mechanisms can be increased. Chapters 6 to 10 show the application of the above-mentioned topics to development of precise vacuum mechanisms with enhanced characteristics. Chapter 6 considers new types of nanoscale precision vacuum mechanisms based on magnetic and electric rheology phenomena. This chapter describes new mechanisms of nanoscale precision invented and developed in Russia using the matrix method presented in Chapter 4. Chapter 7 describes new designs of UHV harmonic rotary-motion mechanisms (this type was invented in the USA) with the best parameters in the world (highest longevity, smallest weight, small outgassing rate, high precision). Chapter 8 describes the set of new types of non-coaxial nut-screw UHV mechanisms invented in the USSR and the method of design of these mechanisms. Chapter 9 shows new types of elastically deformed multi-coordinate mechanisms without friction pairs in vacuum. These mechanisms also were invented in the USSR and have no analogues in the world. Chapter 10 shows methods of estimation and control of microparticle flows generated by vacuum mechanisms.

Before publication, the contents of this book were examined and commented by Professor G.L. Saksaganski (St.-Petersburg, Russia), Professor J.L. de Segovia (Madrid, Spain), Professor T. Sawada (Akita, Japan), Professor K. Nakayama (Tsukuba, Japan), Professor P. Řepa (Praque, Czech Republic), Professor Franek (Vienna, Austria), Dr. M. Sherge (Ilmenay, Germany), and Professor M.J. Furey (USA).

The authors wish to thank all those persons who sent them their high esteem on this work.

With good wishes,
The Authors

About the Authors

Evgueni A. Deulin

Professor Deulin E.A (right) discusses the idea of a new experiment with Professor J.L de Segovia (left).

Evgueni A. Deulin was born in Moscow in 1938. He earned his BSc (engineering) degree at Bauman Moscow State Technical University (BMSTU) in 1962. He earned his first doctoral degree (PhD) with a thesis on the topic of Ultra High Vacuum (UHV) Mechanism Optimal Design in 1977. He earned his second doctoral degree (Dr.Sci) with a treatise on the Theory of UHV Mechanisms Creation in 1987. From 1988 he worked as a professor of Mechanical Engineering Department of BMSTU and lead the student Vacuum Technology Research Group. Twenty-two of his pupils earned a PhD degree during this period.

From 1964–1989 E.A. Deulin researched all main types of UHV mechanisms and designed 54 types of standard (USSR) mechanical vacuum feedthroughs, including precise UHV harmonic rotary feedthrough, long travel UHV linear non-coaxial ones, precise (in nano scale range) multi coordinate magnetic rheology drives etc. He is the author of 46 patents in the USSR and Russia. E.A. Deulin researched the process of gases exchange at friction in vacuum. He was the first, together with Professor J. de Segovia and Dr. R. Nevshoupa, who theoretically (in 1996) described and experimentally researched the mechanically stimulated gases solution process. In 1997 E.A. Deulin was the first to theoretically describe phenomenon of friction force decreasing in UHV for "smooth" surfaces. Today this work has developed into the "Theory of Smooth Surfaces Dry Friction". E.A. Deulin is the author of more than 250 scientific publications. He is the supervisor of a postgraduate students group and he was a winner of title "The Best Student Research Supervisor" at BMSTU.

Yuri V. Panfilov

Yuri V. Panfilov was born in Moscow in 1948. He earned his BSc (engineering) degree in 1972 at Bauman Moscow State Technical University (BMSTU). He earned his first doctoral degree (PhD) with a thesis entitled "Research of Elastomer Feedthroughs for Vacuum Technological Equipement Automation" in 1980. He earned his second doctoral degree (Dr.Sci) in "New Methods of Nanoscale Structures Creation" in 1990. From 1990 onwards he has worked as a professor of Mechanical Engineering Department at BMSTU and is Chair of Mecanical Engineering in Electronics at BMSTY.

During 1972–2005 Y.V. Panfilov researched all main types of UHV equipment for thin films deposition with different new proprties and methods of product parameters control on nanoscale precision and he also created new methods of product defense in vacuum from dust microparticles which may be formed in vacuum. He organized a series of Russian Conferences and Workshops in the area of nanotechnology. He created a number of new methods of nanoscale structures creation on the surface, including the methods of antiwear coverage on the surfaces of different materials. Y.V. Panfilov is the author of more than 200 scientific publications. He is the author of 40 Russian patents. Y.V.Panfilof is president of a Russian scientific society in the area of nanotechnology.

Valery P. Mikhailov

Valery P. Mikhailov was born in Moscow in 1961. He earned his BSc (engineering) degree in 1983 at Bauman Moscow State Technical University (BMSTU). He earned his first doctoral degree (PhD) with work on "Research and Design of Multicoordinate Magnetic Rheology Drive of Submicron Precision for Installations of Electron Beam Lithography" in 1995. He earned his second doctoral degree (Dr.Sci) in "Automatic Control of Positioning Process of Magnetic Rheology Drives of Nanoscale Precision" in 2003. From 1995 he has worked as a lecturer and from 2003 as a professor at the Mechanical Engineering Department at BMSTU.

From 1985–1999 V.P. Mikhailov researched linear travel, non-coaxial, precise (in nanoscale range) multi-coordinate magnetic rheology drives. V.P. Mikhailov is the author of more than 150 scientific publications. He is author of 25 patents in the USSR and Russia. V.P. Mikhailov is the supervisor of the group of postgraduate students which research and create new methods of nanoscale precision drive control. Under his supervision the series of new multicoordinate drives of nanoscale precision was designed and researched. Multicoordinate drives for the main mirrors turning a new large Russian optical telescope of 25 meter diameter were created.

V.P. Mikhailov won the prize for "The Best Student Supervisor and Lecturer" at BMSTU.

Roman A. Nevshupa

Roman A. Nevshupa was born in 1970. He earned his BSc (engineering) degree in 1994 at Bauman Moscow State Technical University (BMSTU) and his first doctoral degree (PhD) in Mechanical Engineering in 1999. From 2003 onwards he works as Associate Professor at the Mechanical Engineering Department of BMSTU.

The research interests of R.A. Nevshupa include vacuum mechanisms, vacuum tribology, tribological coatings in vacuum, mechanically induced desorption and chemical reactions occurring in mechanical contact. R.A. Nevshupa is the author of more than 40 scientific publications. He is the author of three patents. R.A. Nevshupa performed his researches as invited researcher in various laboratories including the Microtribology Laboratory at Ilmenau Technical University, the Laboratory of Surface Physics and Engineering at the Institute of Materials Science of Madrid, Mesotechnology Research Group at the National Institute of Advanced Industrial Science and Technology (Tsukuba, Japan).

Chapter 1
Using Precise Mechanisms in Modern Vacuum Technological Equipment

There are several groups of vacuum technological equipment equipped with vacuum transporting systems. These groups include: thin films coating equipment [1–4], electron and ion lithography [5], molecular epitaxy [6], surface research [7, 8], electron beam equipment for welding, melting, crystal growing; equipment for outgassing, tubeless pumping, assembling vacuum devices (in ultrahigh vacuum up to 10^{-8} Pa) [9]. It is necessary to transfer the samples, heaters, evaporators, screens, sensors into vacuum chambers of this equipment with high to very low speed and high precision. In many cases it is impossible to use any lubricants in vacuum transporting systems because of the special requirements on the residual atmosphere.

The *coating equipment for thin films* is the largest group of vacuum equipment, which uses the precise vacuum mechanisms. In this kind of equipment it is necessary to transfer the wafers or samples into the treatment area (coating, heating, cooling), to move screens of the evaporators and view-ports, to move the evaporators or sensors for measuring thickness of the film.

Every new generation of this equipment, especially equipment for optical film coating, requires a higher degree of vacuum. As it is known, the ever lower residual pressure makes worse the working conditions of the mechanisms in vacuum and decreases the reliability of the mechanisms.

In order to increase the output of the equipment makes it necessary to equip the installations with gate valves, storage devices, load-lock chambers automatic transporting systems.

The thin film coating installation (USSR design YVND-1) with electric arc evaporator and multiple working positions is shown in Figure 1.1. According to the designer's opinion the high output of the installation can be reached owing to the fact that 48 small samples, fixed to and rotating on the mandrels are coated simultaneously. The rotating mandrels fixed on the rotating carousel one by one move to the evaporation zone. To make this process possible the designer used 144 ball bearings (pos. 6, 7) which rotating at 100–300 rps at temperature 200°C in high vacuum

E.A. Deulin et al., *Mechanics and Physics of Precise Vacuum Mechanisms*,
Fluid Mechanics and Its Applications 91, DOI 10.1007/978-90-481-2520-3_1,

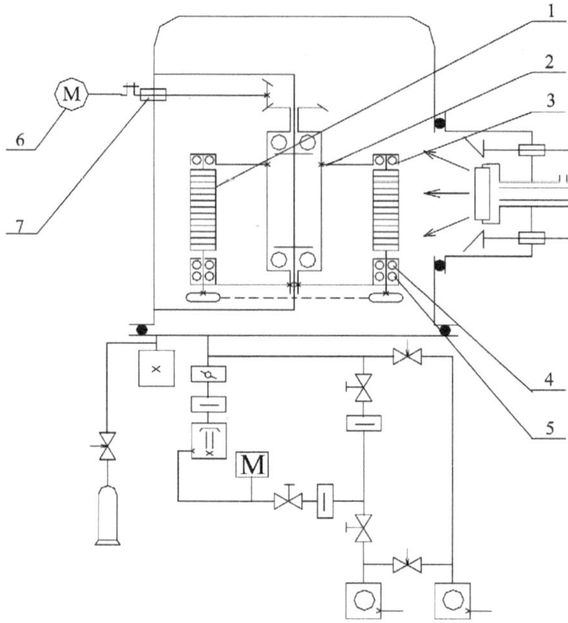

Fig. 1.1 The thin film coating installation with an electric arc evaporator: 1 – cylindrical details assembled on the mandrel; 2 – workpiece carrousel; 3, 4, 5 – ball bearings of the mandrels; 6 – motor, 7 – rotary-motion feedthrough.

(7×10^{-3} Pa). The rotating carousel and mandrels are driven through the rotary-motion with the rotating motion feedthrough (pos. 1). From experience we know that the least reliable component of the installation is a vacuum ball bearings. High output of this installation can be reached only by using a special method of maintenance. The best solution to this problem is using a diagnostic system.

The theory of technological equipment productivity [10] shows that in the case of multiple working positions installation productivity is limited by the idling duration. In the case of vacuum equipment the idling include the pumping time, which for ultrahigh vacuum (UHV) equipment can reach 10–30 hours! To decrease this waste time the designers create installations supplied with load-lock chambers.

The example of UHV installation of thin films coating (UVN-73P-2) [11] is shown in Figure 1.2. The arm of the manipulator 1 carries the samples from the storage drum 2 situated in the load-lock chamber 3 into the working drum 4 situated in the vacuum chamber 5. During the loading operation, the arm 1 has to move through the narrow slots (holes) of the gears 11 of the working drum 2. The precision of the drives 12 and 13 of the carousel and of the working drum, respectively must be quite high to ensure this loading operation. The precision of the arm positioning is about 0.2 mm at the length of its travel 700 mm.

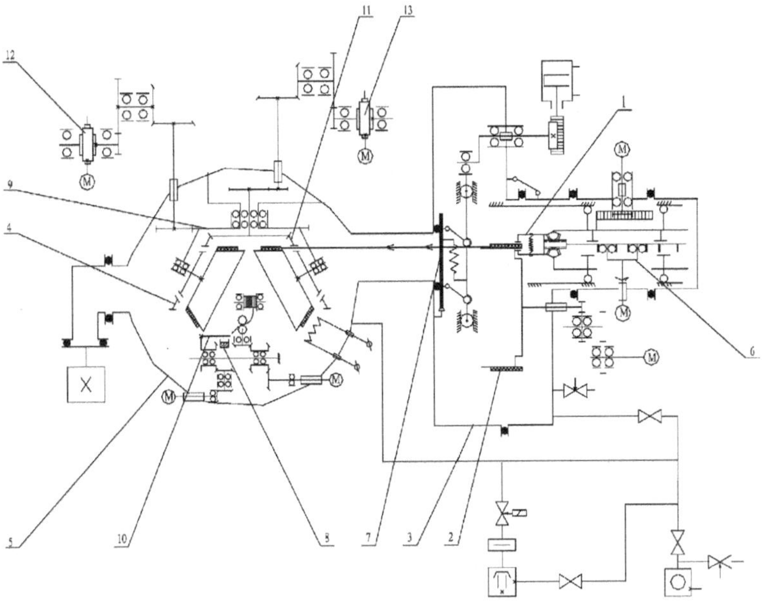

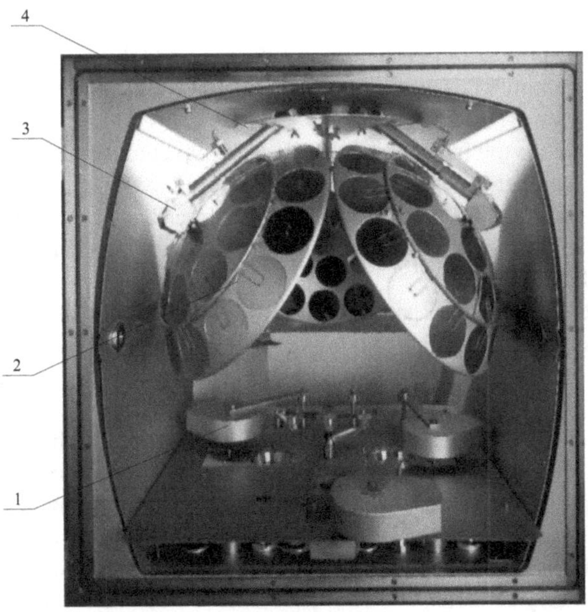

Fig. 1.2 Top: The diagram of the thin film coating installation (USSR design UVN-73P-2): 1 – the arm of the manipulator for the samples loading into vacuum chamber; 2 – storage drum; 3 – sluice chamber; 4 – working drum; 5 – vacuum chamber; 6 – drive of the arm; 7 – gate; 8 – evaporator; 9 – carrousel; 10 – evaporator screen; 11 – gear wheel of working drum; 12, 13 – the drives of the carrousel and the drum. Bottom: The view of the internal vacuum chamber mechanisms of the thin films coating installation manufactured by Balzers Company [1]: 1 – evaporators screens; 2 – working drums, 3 – drums rotation drive; 4 – carrousel.

Table 1.1 The requirements to the mechanisms of thin film coating equipment.

	Parameter	Allowable value
1	Residual working pressure	10^{-6} Pa
2	Outgassing rate of the mechanism	10^{-7} $m^3Pa{\cdot}s^{-1}$
3	Temperature at the outgassing process	700 K
4	Error of linear positioning	0.1 mm
5	Error of angular positioning	0.01 radian
6	Length of the travel	2 m
7	Step of the travel	100–800 mm
8	Transporting speed	10–200 $mm{\cdot}s^{-1}$
9	Rotation speed	1–300 rpm
10	Time of response	0.1 s
11	Axial load	100 N
12	Torque	80 N·m
13	Number of degrees of freedom	3
14	Reliability (the probability of failure before the planned repair)	0.95

The drive of the gate 7 has to provide a normal sealing force of about 500 N. The mechanism of the evaporator 8 has to provide high smoothness of the motion of the wire during evaporation.

The intention of the designers to increase the output of this installation has lead to an increase in the number of the feedthrough mechanisms (up to 8) and ball bearings (up to 50).

Figure 1.2a shows the internal view of the vacuum chamber and the mechanisms of the coating installation for thin films manufactured by Balzers Company [1]. We can see that the internal design of the installation described above closely resembles this one. The carousel 4 with three working drums 2 should provide high productivity of the installation.

According to the theory of productivity [10], we know that as the number of the used mechanisms increases the reliability of the designed installations decreases. So, the requirements for small residual pressure and high output contradict with the requirement of high reliability of the vacuum mechanisms. Table 1.1 shows the main requirements for the mechanisms of coating equipment for thin films.

The *crystal growing equipment* requires very uniform linear motion of the growing crystal in combination with its rotation in high vacuum. Growing speed is very small (0.0001–0.002 ms^{-1}) and the time of manufacturing of one crystal can be three weeks, so the vacuum drivers, pumps and other mechanisms of these installations must be very reliable. Growing processes are usually realized at very high temperature (2000–3000°C), therefore, vacuum mechanisms must be cooled down. The designers usually use a cup-type seal in combination with different types of the drive of the monocrystal transference.

Figure 1.3 shows the kinematic and vacuum schemes of the installation of monocrystal growing with Verneuil's method. The guide screw 10 ensures the vertical transference of the monocrystal 1 being grown. The drive consisting of the motor 6, reducer 7, and coupler 9 ensures the working feed of the monocrystal 1. The mo-

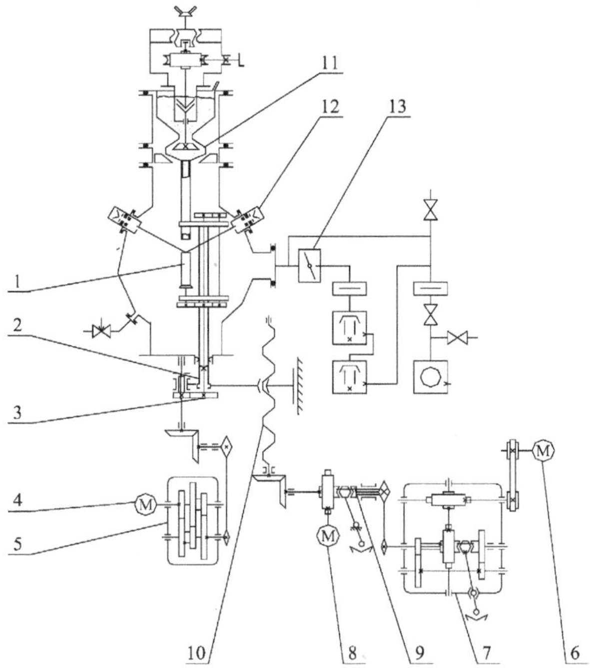

Fig. 1.3 The diagram of the installation of monocrystal growing with Verneuil's method: 1 – the monocrystal, being grown; 2 – the coupling of the vertical feeding; 3 – the drive of the monocrystal rotation; 4 – motor; 5 – reducer of the monocrystal rotation; 6 – motor; 7 – reducer of vertical transference (feeding); 8 – motor; 9 – socket of fast monocrystal transference; 10 – the guide screw of the monocrystal feeding; 11 – choker; 12 – electron gun; 13 – the gate of the vacuum system.

tor 8 ensures the fast monocrystal transference. The drive, consisting of motor 4 and reducer 5 ensures the rotation of the monocrystal 1 being grown. The linearly moving hollow spindle 2 and rotating shaft 3 are sealed with two coaxial cup-type seals. The choker 11 has to ensure the stability of the working dust feeding to the melting zone. The drive of the choker is also sealed with the cup-type seal. The melting instrument consists of two electron guns 12. The pumping system pumps the air out of the work chamber through the gate 13 till the pressure reaches 10^{-3} Pa.

The routine method of monocrystal growing, which is known as the Czochralski technique, is illustrated in Figure 1.4. This method sets high requirements on the mechanism of the touch-string monocrystal transference: speed of vertical transference is 0.0001–0.002 mm$\times$s^{-1}, length of the travel is about 500 mm, speed instability is less than 0.1%, speed of monocrystal rotation is 40 rpm. The diagram 1.4 shows that for the working (slow) monocrystal (pos. 1) vertical transference the harmonic gear reducer 2 was used in combination with nut-screw pair 3. The drive 4 is used for a fast touch-string manual transference. The drive 5 is used for the touch-string rotation.

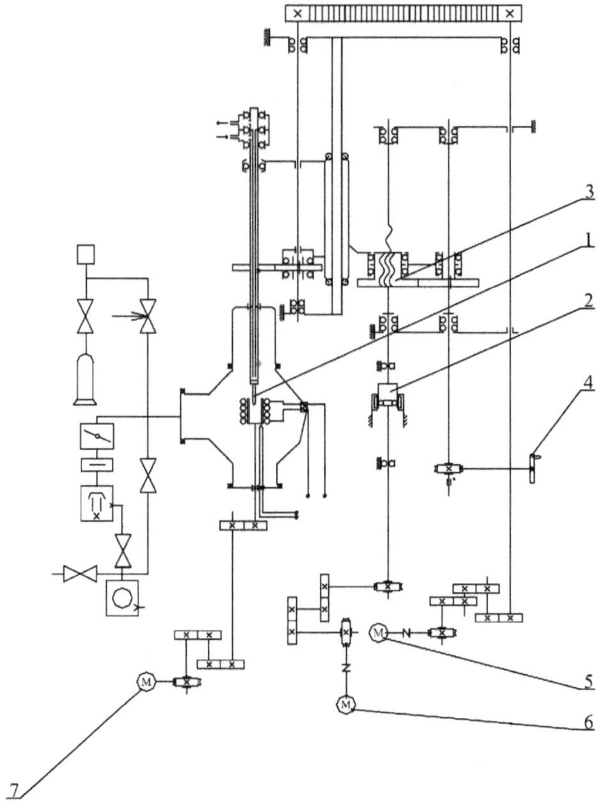

Fig. 1.4 The diagram of the monocrystal grow installation based on the Czockhralski method: 1 – touch-string of a monocrystal; 2 – harmonic drive for the monocrystal touch-string transference; 3 – nut-screw drive; 4 – drive of the fast touch-string transference; 5 – drive of the touch-string rotation; 6 – motor of the touch-string transference; 7 – motor of the touch-string rotation.

Table 1.2 The requirements to the mechanisms of crystal growing equipment.

	Parameter	Allowable value
1	Residual working pressure	10^{-5} Pa
2	Length of the travel	0.8 m
3	Transporting speed	10^{-4} mm·s^{-1}
4	Softness (uniformity) of transference	0.1%
5	Reliability (the probability of failure less work till planned repair)	0.99

So, we can see that the requirements of small residual pressure in combination with high temperatures are in conflict with the requirements of high reliability of the vacuum mechanisms. Table 1.2 shows the main requirements to the mechanisms of crystal growing equipment.

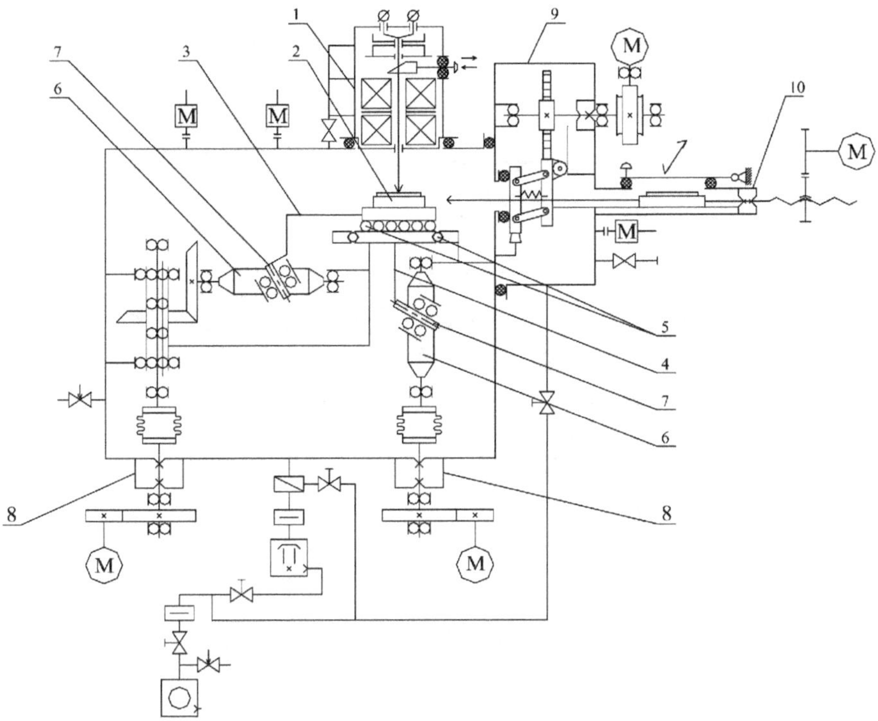

Fig. 1.5 The diagram of the electron beam lithography installation with friction drive of the table: 1 – electron beam focusing system; 2 – positioning table; 3 – drive of linear transference of the table; 4 – drive of the cross transference of the table; 5 – ball guides of the table; 6 – driving shafts of the table; 7 – inclined rolls of the table drive; 8 – rotary-motion feedthroughs; 9 – sluice; 10 – rotary-motion feedthrough of the feeder.

The *equipment of electron, X-ray and ion lithography* allows us to realize processes of direct formation of microstructures on clean surfaces in vacuum. Small range of electron or ion beam deviation (1–10 microns) and impossibility of X-ray deviation require use of precise vacuum mechanical scanning systems. Usually it is two-coordinate drives which work in indexing or continuous motion regimes in high vacuum (in case of electron or ion lithography) or in protective gas (in case of X-ray lithography). These drives must ensure high precision of positioning (error less then 0.5 micron) in the "start-stop" regime of the sample transference at high speed of the scanning (till 200 mm s^{-1}).

Figure 1.5 shows the kinematic and vacuum schemes of the electron beam lithography installation with friction drive of the worktable. The electron beam focusing system 1 forms the electron beam, which forms the required topology on the sample fixed on the positioning table 2. The positioning table is based on the ball guides 5 and is driven with two orthogonal drives 3 and 4. The original inclined rolls 7 in combination with driving shafts 6 are used for the table precise positioning. The

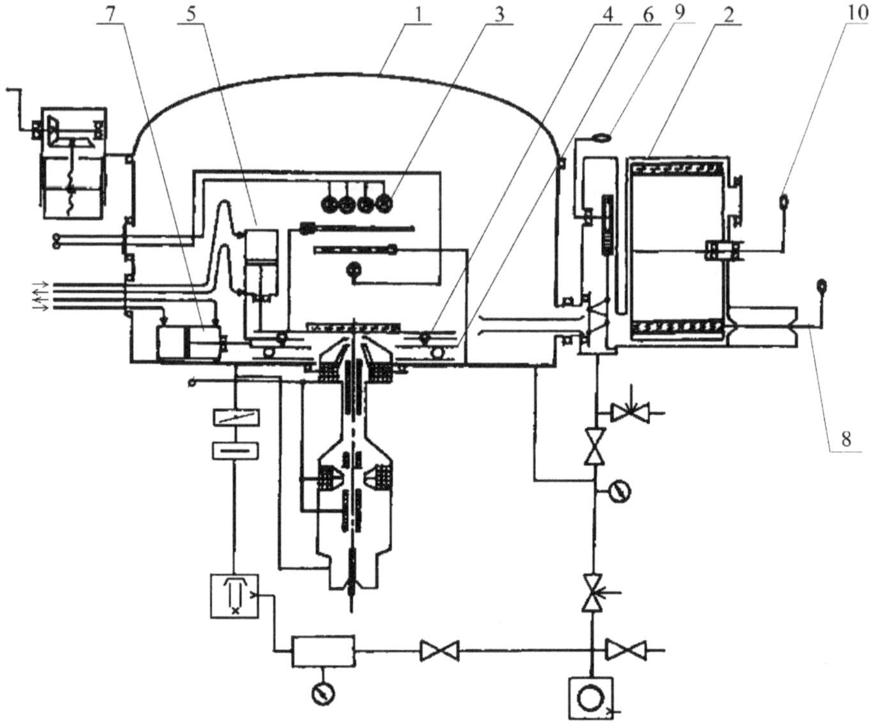

Fig. 1.6 The diagram of the electron beam lithography installation based on a hydro drive: 1 – work chamber; 2 – sluice chamber; 3 – light-emitting diodes of raster coordinate counting system; 4 – cross pilot-bearing of the coordinate table; 5 – hydro drive of cross transference; 6 – pilot-bearing of the coordinate table; 7 – hydro drive of the coordinate table transference; 8 – manual drive of the samples feeder; 9 – drive of the gate; 10 – drive of the storage drum.

precision of positioning is about 1 micron. The electron beam inclination ensures fixation the final element. The speed of the table linear transference is 50 mm s^{-1}. The load-lock chamber 9 with loading device 10 is used to shorten the pumping process and to increase the output of the installation.

Figure 1.6 shows the kinematic and vacuum schemes of another design of electron beam lithography installation. The high productivity in this case [5] is reached using the hydro-drive of a worktable situated in the work chamber 1. The load-lock chamber 2 helps to shorten the pumping time during the loading process. The positioning system includes light-emitting diodes 3, photo diodes and raster coordinate counting system. The work coordinate table is fixed on cross pilot-bearings 4 and 6 and is driven with cross hydro drives 5 and 7.

The automatic drive of the worktable must ensure the precision of the table positioning about 1 micron and fast action 0.005 sec. The requirement of high fast action contradicts in this case with the requirement of high precision in high vac-

Table 1.3 The requirements to the mechanisms of the equipment of electron, X-ray and ion lithography.

	Parameter	Allowable value
1	Residual working pressure	10^{-6} Pa
2	Length of the travel	200 mm
3	Range of the step	1–10 mm
4	Transporting speed	200 mm·s^{-1}
5	Operating speed (time of response)	0.005 s
6	Error of positioning	0.5 micron
7	Error of element stitching	0.1 micron

uum. High speeds of the drive cause oil evaporation into residual vacuum because the hydrodrive is placed inside the vacuum chamber.

Now, the designers attempt to ensure oilless UHV in the vacuum chamber and in such a way to increase the longevity of the electron-optical system. However, these attempts were not very successful because of the contradicting requirements of constant growth. Table 1.3 shows the main requirements to the mechanisms of equipment of electron and ion lithography.

The *equipment for assembling vacuum devices* exists in the form of the equipment of electron-beam welding, equipment of high frequency current soldering and equipment of assembling vacuum devices and cold welding. The *equipment of electron-beam welding* and the *equipment of high frequency current soldering* use high vacuum, about 10^{-4} Pa, because this pressure ensures enough length for traveling of the electrons in an electron beam and also sufficient cleanness of the surfaces being joined at processes of welding or soldering.

Figure 1.7 shows the kinematic and vacuum schemes of the electron beam welding installation of the carousel type. Electron gun 1 welds the components fixed to the spindle 2. Drive 3 of the spindle vertical transference takes the detail to the work zone. Rotary-motion feedthrough 4 with cup-type seal is used for spindle rotation. Cross wheel 6 is used for periodical carousel turn. The shaft of the carousel is sealed by a cup-type seal 7. The nut-screw pairs 9 and 10 are used for adjusting the electron gun. The bellow 8 plays the role of linear motion feedthrough. The ultimate residual pressure is limited because of the use of the cup-type seal.

Figure 1.8 shows the kinematic and vacuum schemes of the installation of ultra-high frequency (UHF) soldering in vacuum. The installation contains three similar drives, every one of which contains motor 1, reduction gear 2, bellows sealed rotary-motion feedthrough 3. The left one of these drives is used for the spindle 4 vertical transference on the length higher than 500 mm. The middle one is used for the periodical carousel 5 rotations. The right one is used for the periodical carousel horizontal transference by the nut-screw pair driven by gear 8. The carousel is shown in its left position. After the carousel transference has reached its right position, the spindle 4 rises about 50 mm upwards and its gear 7 comes into coupling with gear 5 of the carousel. Then the middle drive turns the carousel in the next position and

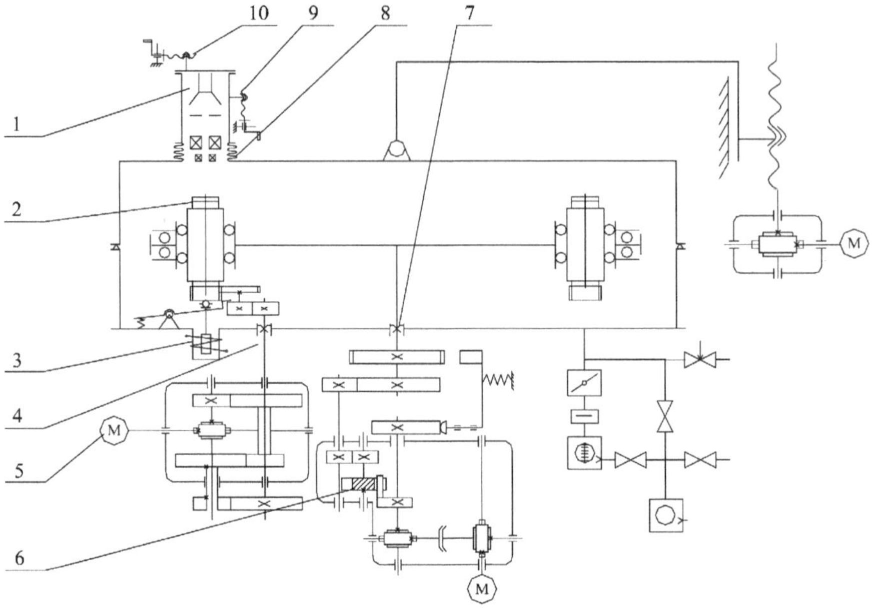

Fig. 1.7 The diagram of the electron beam welding carousel installation: 1 – electron gun; 2 – spindle with the detail being worked; 3 – drive of the spindle vertical transference; 4 – rotary-motion feedthrough; 5 – motor of the spindle rotation; 6 – cross wheel for periodical carousel turning (rotation); 7 – rotary-motion feedthrough.

the left drive raises the component being soldered to the upper position (inside UHF coil 6).

The large number of different kinematic pairs reduces the reliability of this installation. The bellows sealed rotary-motion feedthrough 3 ensure precision at small angles (0.01–0.02 radian) and is more suitable for use in an oilless ultrahigh vacuum applications.

The *equipment for assembling vacuum devices and cold welding* use the ultrahigh vacuum which ensures good emission ability for welded vacuum gauges. In case of night vision 3 assembling the working pressure in vacuum chamber must be about 5×10^{-9} Pa.

The length of the travel of the transporting mechanisms must allow the transference of a welded component (cold cathode in case of night vision gauge) from one vacuum chamber to another one through the vacuum gate. This distance can reach 1 meter. The precision of transporting mechanisms must ensure precise positioning of the component being welded in ultrahigh vacuum.

The kinematic and vacuum scheme of the simplest installation of photoelectron gauge assembling in vacuum is shown in Figure 1.9. The photo-cathode is activated by the radiation passing through the window 12 into the work chamber 1 at work pressure 5×10^{-9} Pa. The linear motion feedthrough 4 transfers the activated cath-

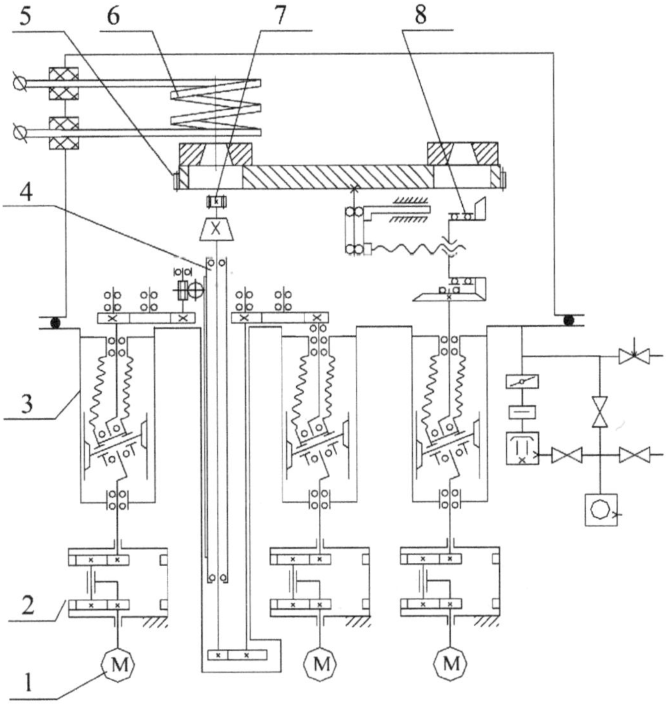

Fig. 1.8 The diagram of the installation of ultrahigh frequency (UHF) soldering in vacuum: 1 – motor of the rotary-motion feedthrough; 2 – reduction gear; 3 – bellow sealed rotary-motion feedthrough; 4 – spindle for the detail being worked vertical transference and rotation; 5 – gear for the carousel rotation (now carousel is shown in its left position); 6 – the coil of UHF inductor; 7 – spindle gear (coupling with gear 5 at carousel rotation in the right position of the carousel); 8 – drive of the carousel transference (left–right).

ode into the photoelectron gauge 2 being assembled through stem 3 of the gauge. To ensure a good quality of the photo-cathode, the spectrum of the residual gases should not contain any carbohydrates, oxygen, water vapors and methane. The resulting outgassing rate has to be lower than 10^{-8} m^3 Pa s^{-1}. The maximum temperature of outgassing is 400°C. The final pumping is done with vacuum pumps 7, 8 and cryogenic adsorption pumps 9, 10. The rough pumping is done with rough-vacuum pump 11 and adsorption pumps 9, 10.

The requirements to the linear-motion feedthrough are the following: photo-cathode positioning against the window 12 with precision 0.1 mm; the stock of the feedthrough 8 deviation below 0.2 mm; the axial load on the stock of the feedthrough higher than 400 N at the gauge assembling.

Table 1.4 shows the main requirements to the mechanisms of the equipment of vacuum devices assembling and cold welding.

The *equipment of molecular beam epitaxy* is designed for the growing of thin monocrystal films with the use of a set of molecular or atomic beam sources. The

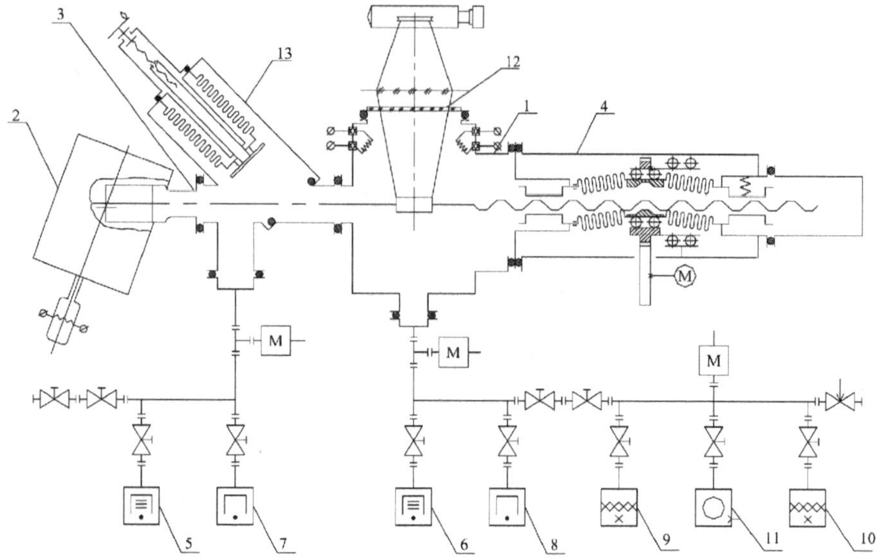

Fig. 1.9 The diagram of the installation of photoelectron gauge assembling in vacuum: 1 – work chamber for photo-cathode forming; 2 – the photoelectron gauge being assembled; 3 – stem of the photoelectron gauge; 4 – linear motion feedthrough; 5, 6 – cryogenic sorption pumps; 7, 8 – vacion pumps; 9, 10 – adsorption pumps; 11 – rough-vacuum pump.

Table 1.4 The requirements to the mechanisms of the equipment of vacuum devices assembling and cold welding.

	Parameter	Allowable value
1	Residual working pressure	$5 * 10^{-9}$ Pa
2	Outgassing flow from the mechanism	10^{-9} $m^3 \cdot Pa \cdot s^{-1}$
3	Length of the travel	1 m
4	Error of positioning	0.2 mm
5	Axial load in welding mechanism	100–1000 N
6	Reliability (the probability of failure less work till planned repair)	0.99

main condition of thin monocrystal films growth is the cleanness of the wafer surface during the coating process which could last 10–30 hours. To ensure such cleanness, the gas pressure in installation must be about 10^{-9} to 10^{-10} Pa. The time to reach such ultrahigh vacuum could be from several days to 1–2 weeks. It means that the reliability of this equipment including the vacuum mechanisms must be very high.

The temperature of outgassing heating can be up to 450°C. Ultrahigh vacuum in combination with high temperature reduces the reliability of the mechanisms and it contradicts with the requirement of high reliability of the equipment.

Figure 1.10 shows kinematic and vacuum scheme of the simplest installation of molecular beam epitaxy. The chambers of the evaporators 1, 2, 3 contain the certain

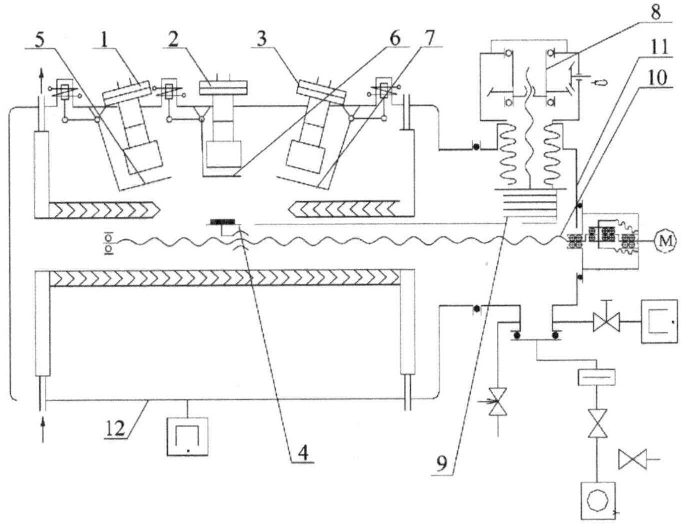

Fig. 1.10 The diagram of the installation of molecular beam epitaxy: 1, 2, 3 – evaporators; 4 – the carrier with the sample; 5, 6, 7 – the screens of the evaporators; 8 – linear motion feedthrough for the carrier transference; 9 – the samples magazine; 10 – the carrier drive; 11 – sluice chamber.

Table 1.5 The requirements to the mechanisms of the equipment of molecular beam epitaxy.

	Parameter	Allowable value
1	Residual working pressure	$5 * 10^{-10}$ Pa
2	Outgassing flow from the mechanism	10^{-10} $m^3 \cdot Pa \cdot s^{-1}$
3	Outgassing temperature	450°C
4	Error of positioning	0.2 mm
5	Number of degrees of freedom of transporting mechanism	3–4
6	Reliability (the probability of failure less work till planned repair)	0.999

materials coated on the sample 4. The screens 5, 6, 7 are used for control of the flows of the materials being evaporated. The linear motion drive 7 transfers the container 9 with the stored samples in vertical direction. Linear-motion feedthrough 10 transfers the carrier with the sample 4 from the container 9, situated in the load-lock chamber 11 to the work chamber 12. The total outgassing rate has to be below $10^{-9}m^3$ Pa s^{-1}. The vacuum chamber is pumped by vacuum pumps and cryogenic sorption pumps during 1–2 weeks. This installation has to work without repair during 0.5–1.5 years and the requirement of its reliability are unusually high.

Table 1.5 shows the main requirements to the mechanisms of the equipment of molecular beam epitaxy.

It is obvious that the way to eliminate these contradictions is to design new mechanisms based on new principles. These mechanisms must be of high reliability. The

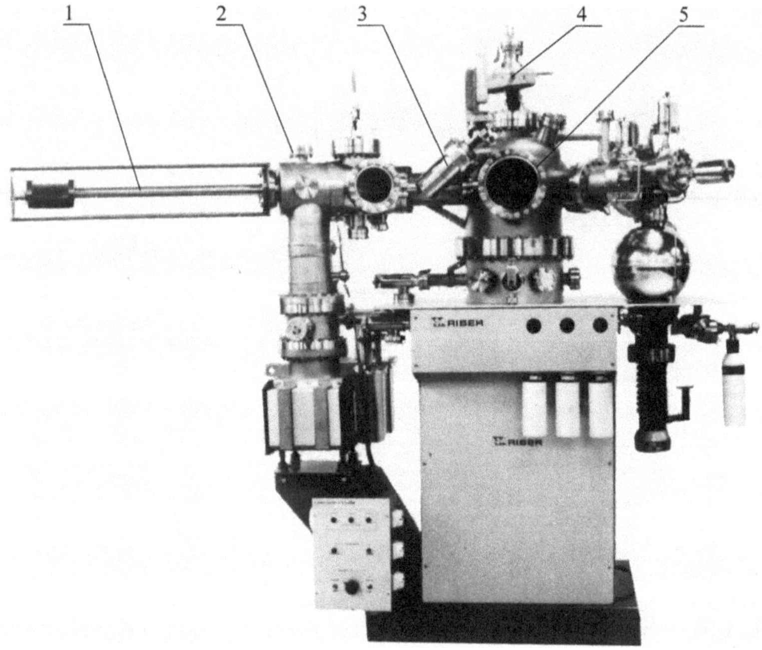

Fig. 1.11 The general view of the analytical installation of Riber Co. [8]: 1 – two degrees of freedom magnet vacuum manipulator; 2 – sluice chamber; 3 – inlet vacuum gate valve (inclined design); 4 – positioning vacuum manipulator; 5 – work chamber.

other way is the developing of better diagnostic methods as well as the rules of maintenance and repair.

The *surface research equipment* enables a number of processes which can protect the effective electronics devices appearance/scanning.

These processes include:

- the uniformity of surface analysis;
- the impurity or doped level analysis;
- the impurity or doped level distribution analysis;
- the surface structure analysis.

Most methods work by analysis of the secondary particles which appear as a result of the bombardment of the surface by primary particles. For the practical realization of this principle, it is necessary to use a drive which realizes the sample positioning with standard error of positioning the characteristic dimensions of the working primary beam. Usually these dimensions: d – diameter of the primary beam crossover and h – definition in depth are ranged betwen 10 and 1 microns.

All methods of surface analysis are very sensitive to the cleanness of the surface which directly depends on vacuum degree. It is known that the time of clean surface conservation: $T_a = 4 \times 10^{-4} \times p^{-1}$ s, where p is the residual pressure, Pa. So, at

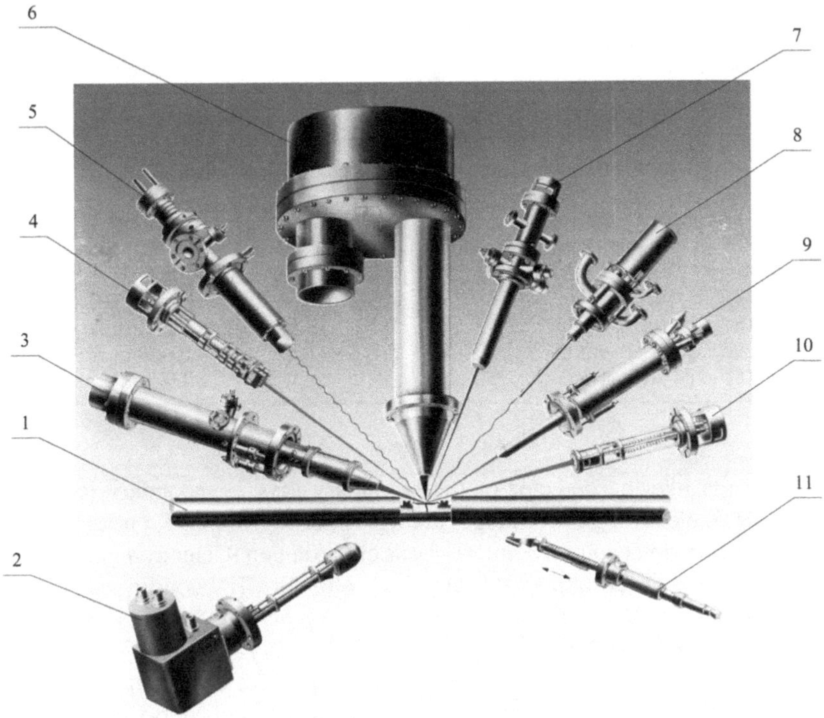

Fig. 1.12 The scheme of the research and technology instruments positioning [7] in vacuum about the searched wafer: 1 – searched wafer; 2 – detector of the secondary ions; 3 – secondary ions mass-spectrometer; 4 – electron gun for Auger analysis; 5 – X-ray source; 6 – energy analyzer; 7 – ion gun; 8 – ultraviolet source; 9 – micro focus electron gun; 10 – electron gun; 11 – Faraday cup.

pressure 10^{-7} Pa this time is only about 1 hour, which is too short for the analysis. Usually, the surface research equipment requires a working pressure of 10^{-8} Pa.

The general view of the analytical installation of Riber Co. [8] for LEED, AUGER, ESCA, SIMS analysis is shown in Figure 1.11. The analyzed samples are transferred by the manipulator 1 from the load-lock chamber 2 through the inlet vacuum valve 3 to the work (analytical) chamber 5, which is pumped with UHV pumping system.

Different analytical instruments include SIMS analyzer, AES electron gun, X-RAY gun, energy analyzer, ION gun, ultraviolet source, SAM Microfocus electron gun, flood gun, Faraday cup, and others. These instruments work properly if the analyzed surface is kept in certain positions at high precision in respect to the instrument.

The scheme of the research and technology instruments positioning [7] in vacuum above the analyzed wafer 1 is shown in Figure 1.12. We can see that differ-

Table 1.6 The requirements for the mechanisms of the surface research equipment.

	Parameter	Allowable value
1	Residual working pressure	10^{-9} Pa
2	Number of degrees of freedom of transporting mechanism	6
3	Length of the travel along X, Y axes,	40 mm
	along Z axis	20 mm
4	Error of linear positioning	1 micron
5	Range of angular rotation of a sample around X axis;	360°
	around Y axis;	100°
	around Z axis	360°
6	The accuracy of angular readout	1 angular sec
7	The angular drive accuracy	10–50 microns
	The axial drive accuracy	0.5–1 microns

ent instruments like the detector of the secondary ions 2, secondary ions mass-spectrometer 3, electron gun for Auger analysis 4, X-ray source 5, energy analyzer 6, ion gun 7, ultraviolet source 8, micro-focus electron gun 9, electron gun 10, Faraday cup 11 require the certain and precise orientation in UHV conditions.

Table 1.6 shows the main requirements for the mechanisms of the surface research equipment.

The *equipment of vacuum scanning microscopy* enables observing surface morphology with atomic resolution, measuring atomic forces and energy, analyzing the structure, topology and chemical variations in the composition without distortions caused by the adsorbed gases. The general view of high vacuum scanning microscope is shown in Figure 1.13a. The internal mechanisms of the microscope are mounted on a vacuum flange as shown in Figure 1.13b. The principle of operation of a scanning microscope is shown in Figure 1.14. The piezo drive 7 (piezo tube) ensures scanning motion of the sample 6 in the range 3–5 microns along the X, Y coordinate axes. The probe 5 deflection which corresponds to the profile of the sample 6 variation is measured using the input receiver 3 and is then sent to computer 1 through the converter 2. It is clear that the adsorbed gas film of one monolayer (0.3 nanometer) is able to distort the results of the measurement. All methods of scanning vacuum microscopy are very sensitive to the cleanness of the surface. That is the reason for using ultrahigh vacuum instruments.

Usually, different instruments of vacuum scanning microscopy, for example Scanning Tunneling Microscopy (STM), Atomic Force Microscopy (AFM) among others, are joined in one ultrahigh vacuum installation.

To ensure the required cleanness, the working vacuum in the equipment of vacuum scanning microscopy must be about 10^{-9} to 10^{-10} Pa.

The required accuracy and the minimal scanning resolution of the scanning mechanisms must correspond to the dimensions of one atom. The existing piezo mechanisms ensure scanning areas about 4×4 microns^2 with a resolution of 0.1 nanometer. This scanning area is too small for the solution of the modern tasks of nanotechnology and it must be increased to the size about 1×1 mm. This task can

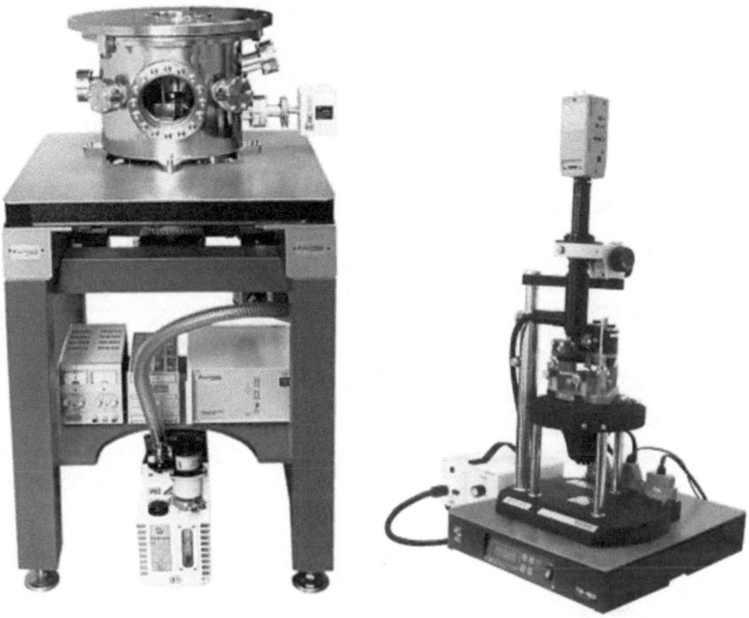

Fig. 1.13 The general view of high vacuum scanning microscope (a) and view of its internal vacuum mechanisms (b) mounted on a flange.

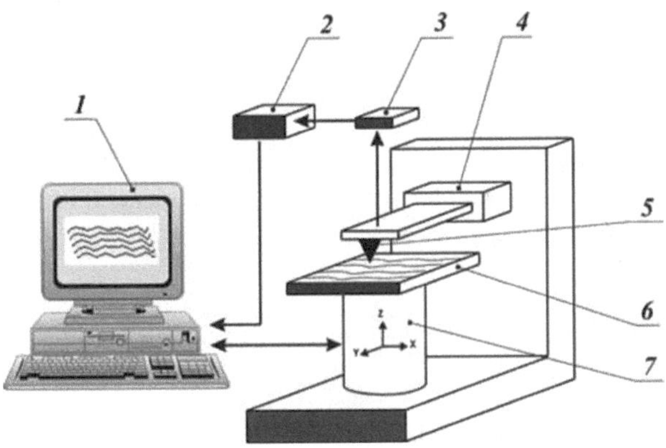

Fig. 1.14 The general scheme of scanning tunneling microscope (a): 1 – computer, 2 – converter, 3 – input receiver (of tunneling current or cantilever deflection), 4 – probe holder, 5 – probe, 6 – researched sample, 7 – piezo ceramic drive; and the scheme of tunneling microscopy (b): 1 – probe (W filament), 2 – the researched surface.

be solved using a secondary drive. The linear resolution of this secondary drive must be about 0.3 nm.

Table 1.7 shows the main requirements for the mechanisms of vacuum scanning microscopy equipment.

Table 1.7 The requirements for the mechanisms of vacuum scanning microscopy equipment.

	Parameter	Allowable value
1	Residual working pressure	10^{-9} Pa
2	Number of degrees of freedom of transporting mechanism	3
3	Length of the travel along X, Y axes,	1 mm
	along Z axis	1 micron
4	Error of linear positioning along X, Y axes	0.3 nm

Table 1.8 The main requirements for the mechanisms of vacuum technological equipment.

	Parameter	Allowable value
1	Residual working pressure	10^{-9} Pa
2	Outgassing flow of the mechanism	10^{-10} m^{3}Pa·s^{-1}
3	Temperature at the outgassing process	750 K
4	Error of linear positioning	0.3 nm–0.1 micron
5	Error of angular positioning	30 angular sec
6	Length of the travel	2 m
7	Step of the travel	0.8–10 mm
8	Transporting speed	0.0004–300 mm·s^{-1}
9	Rotation speed	0.1–300 rpm
10	Time of response	0.05 s
11	Axial load	100–1000N
12	Torque	80 N·m
13	Number of degrees of freedom	6
14	Reliability (the probability of failure less work till planned repair)	0.8–0.999

We have considered the different groups of vacuum equipment in order to decrease the tolerances of the different vacuum mechanisms.

Table 1.8 shows the main requirements to the mechanisms of vacuum equipment.

References

1. Booklet of "VAT Valves Product Line", BPRL91E1, 1991, 12 pp.
2. Booklet of "Balzers", The Balzers Group, BA800074PE (8801), Liechtenstein, 1987, 16 pp.
3. Booklet of "Leybold": Linear and/or Circular Coating System, 012.8.60.52.037.02, Hanau (Germany), 1988, 4 pp.
4. Booklet of "ULVAC": High Vacuum Equipment and System, E0203-1, 90101000ADP, Tokyo, 20 pp.
5. Pancratov I.V., Satarov G.K., Deulin E.A., et al., *Analysis of Productivity of Electron/Ion-Lithography Equipement*, Electronika, Moscow, 1986, 44 pp. [in Russian].
6. Booklet of "Riber", MBE 608.15H82.710, 1988, 16 pp.
7. Booklet of "Leybold–Heraeus": Leybold–Heraeus in Surface Analysis and Laser Mass-Spectrometry. A Short Portrait, 190-1000.2/2494, Köln, 1985, 8 pp.
8. Booklet of "Riber": 608.40j595, Ruel, Malmaison (France), 1988, 803 pp.
9. Kostin A.B., Filimonova G.A., Technological installations for photoelectron tubes manufacturing with usage of transference method, *Electronica* **15**, 1983, pp. 217–255 [in Russian].

10. Shaumian G.A., *The Complex Automation of the Technological Process*, Mashinostroenie, Moscow, 1973, 640 pp.
11. Deulin E.A., *Mechanics and Physics of Precise Vacuum Mechanisms*, Vol. 1, Vladimir State University, 2001, 176 pp. [in Russian].

Chapter 2
Typical Vacuum Mechanisms

2.1 Functions of Vacuum Mechanisms

The vacuum mechanisms are used in vacuum technological equipment described in Chapter 1 for transportation of components (samples, wafers, cathodes, etc.) from one working position to another, for their orientation and positioning against the instruments (electron-, ion-, molecular, X-ray beam), or for their positioning against other components (for example, cathode against anode of the gauge). In scanning microscopy they are used for the instrument (probe or cantilever) positioning and scanning of the analyzed surface. In most cases it is enough to have one or two degrees of freedom (linear motion and rotation). However, some technological processes (surface analysis, scanning nanomicroscopy, X-ray lithography) require 3–6 degrees of freedom.

The main element in any vacuum mechanism is the motion feedthrough which plays a role of the mechanical element of transmission as well as a sealing element. It is the most critical mechanical element of the kinematic chain of the drive because it controls the required precision of the mechanism and at the same time it should protect vacuum elements against the atmospheric air. In many cases the motion feedthrough plays a role of the final vacuum mechanism on a technological position.

In some cases for small micro- or nanotravels independent piezo drive situated in the vacuum chamber can be used. In these cases vacuum feedthroughs are not used.

2.2 Rotary-Motion Feedthroughs

Rotary-motion feedthroughs for transmission of the rotation into a vacuum chamber without reduction are well-proven mechanical devices. In high vacuum (to 10^{-5} Pa) they are designed on the basis of a cup-type seal [1, 2] which, in turn, is based on Wilson seal [3]. For rotary-motion transmission into ultrahigh vacuum bellows

E.A. Deulin et al., *Mechanics and Physics of Precise Vacuum Mechanisms*, Fluid Mechanics and Its Applications 91, DOI 10.1007/978-90-481-2520-3_2,

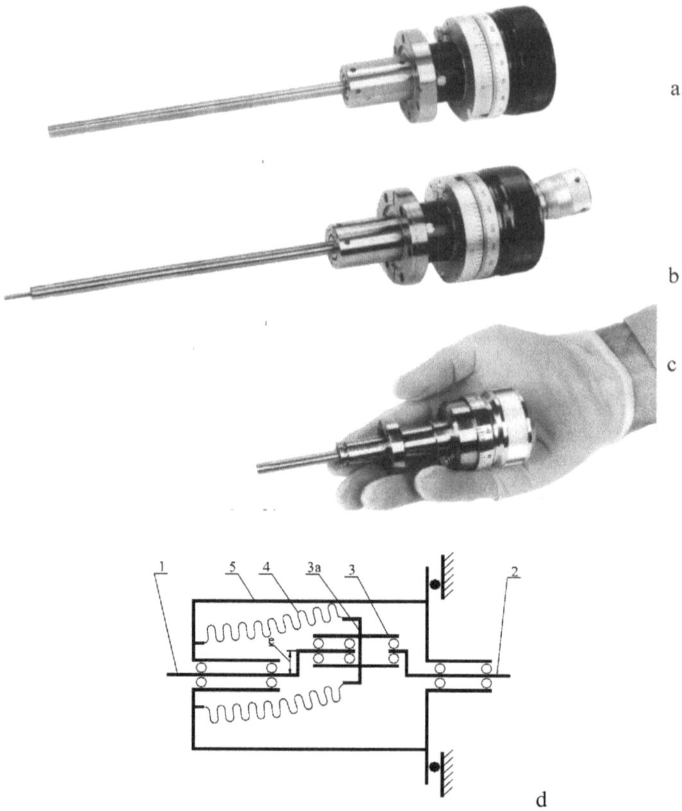

Fig. 2.1 The general view of the set of the bellow sealed rotary-motion feedthroughs (a, b, c), which are manufactured by Vacuum Generators Co. The middle sample (b) is of two degrees of freedom; (d) the diagram of the bellow sealed rotary-motion feedthrough, where: 1 – leading eccentric shaft; 2 – the output eccentric shaft; 3 – intermediate hermetic bush; 3a – the hermetic screen into the bush; 4 – the stainless bellow.

sealed rotary-motion feedthroughs are used. These devices are manufactured by a number of vacuum companies [4–11].

Figures 2.1a–c show the bellows-sealed rotary-motion feedthrough manufactured by Vacuum Generators Co. [9]. These feedthroughs can be designed with double coaxial input and output shafts (Figure 2.1d) which allows both rotation and linear motion, Figure 2.1b. The scheme of the simplest rotary-motion feedthrough is shown in Figure 2.1d. The input rotation of the leading eccentric shaft 1 sets the intermediate hermetic bush 3 in circular motion without rotation. This bush in its turn sets the output eccentric shaft 2 into rotation again. The bellow 4 hermetically welded to the body 5 and to the bush 3 is sealed with the hermetic screen 3a into the bush. From one side, the value of the eccentricity "e" of the shafts 1 and 2 determines the bellows longevity from one side and from the other side the loading ability of the feedthrough from another side. To ensure the maximal longevity of the bellows, the eccentricity must be minimal and, *vice versa*, the eccentricity must be maximal to ensure minimal longevity (minimal radial load) of the ball bearings. The precision and the angular rigidity of the feedthrough increases. Thus, the requirements of the feedthrough contradict one to another. The designer's task is to find the optimal balance between the output parameters and to choose the optimal eccentricity of the shafts.

The harmonic gear rotary-motion feedthrough [12–15] is the precise drive and reducer and the sealing ultrahigh vacuum element simultaneously. The principle of this feedthrough is such that it is able to solve the contradicting requirements. Its parameters are very close to the requirements, shown in Tables 1.2 and 1.6. Therefore, this mechanism is very promising for use in the equipment of surface analysis and crystal growth.

Figure 2.2 shows the general view (a) and the scheme (b) of the harmonic gear rotary-motion feedthrough [12, 13]. The input rotation of the leading shaft 1 and of eccentrically fixed ball bearings transforms motion into circular wave deformation of the flexible gear 3 using the eccentrically fixed ball bearings 2 and 2a, which form the "wave generator". The flexible gear is hermetically welded to the body 6, so, the output shaft 5 based on the ball bearings 7 and 7a is situated in the vacuum zone.

The rotation of the input shaft 1 together with the wave generator (ball bearings 2, 3, fixed on the eccentrics) causes the rotation of the deformation of the flexible gear with the speed of rotation $2n$ (where n is input shaft speed rotation). The flexible gear is fixed to the body 6 and as a consequence of the teeth number difference z_1–z_2 it rotates with the speed of rotation $n2 = n1 \cdot i$. The deformation of the flexible gear 3 (thin-wall sealing element and gear simultaneously) ensures the coupling of the flexible gear 3 and rigid gear 4 only in zones of maximum deformation.

The teeth number of the gears is large (200–300), but the relative deformation of the flexible gear is small (2.5–5.0 tooth module). As a consequence, the teeth number difference is also small. As a result, this type of feedthrough ensures a large gear ratio $i = 80$–300 which can be calculated:

$$i = \frac{z_2 - z_1}{z_2},$$

where z_1 is the number of teeth of flexible gear (of thin-wall sealing element); z_2 is the number of teeth of rigid gear which is fixed on output shaft.

The small thickness of the flexible gear (0.3 mm) and its small radial deformation ensure its high longevity. The external view of the flexible and rigid gears can be seen in Figures 2.2c and 2.2d.

Two dimension-type of the harmonic rotary motion feedthrough (type "BBB-100-40-72" [12, 13]) shown in Figure 2.2b ensures a large ratio ($I = 84$–164) and small kinematic error (less than $6'$).

The main advantages of this type of feedthrough are the following: large gear ratio in one device, high rigidity, small wear of teeth and ball bearings, all-metal design and high allowable temperature of outgassing.

The detailed analysis of the design and the discussion of the features of these feedthroughs are presented in Chapter 7.

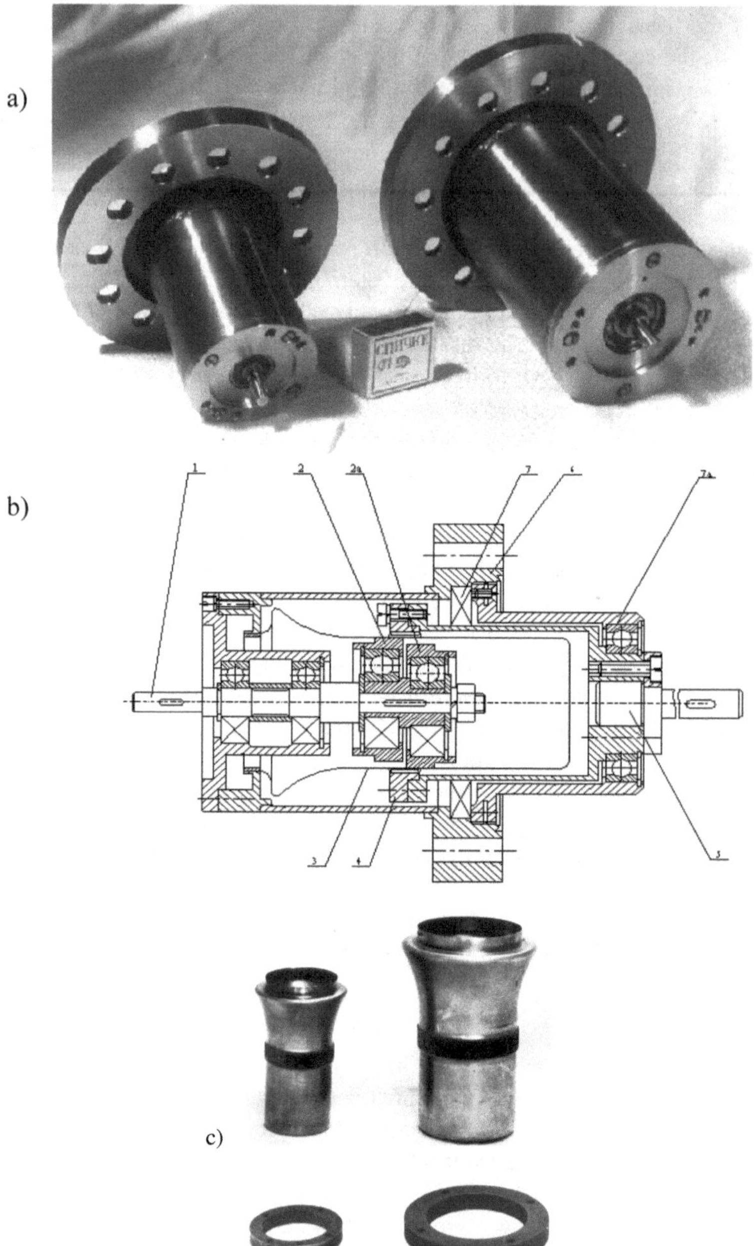

Fig. 2.2 Harmonic gear rotary motion feedthrough. (a) General view of two standard sizes of feedthrough. (b) Scheme of the design; the figures denote: 1 – leading shaft; 2, 2a – rings of the wave generator; 3 – flexible gear (thin-wall sealing element and gear simultaneously); 4 – rigid gear; 5 – output shaft; 6 – body; 7, 7a – ball bearings of output shaft. (c) The general view of two dimension-type of dimension of flexible gears. (d) The general view of two dimension-type of rigid gears.

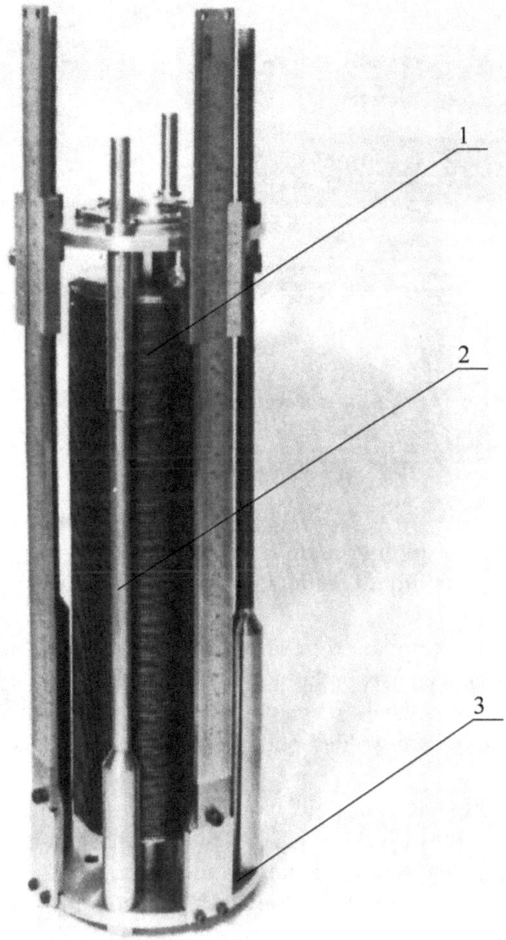

Fig. 2.3 The linear motion feedthrough based on the arc-welded bellow designed by Leisk Co. [16]: 1 – arc-welded bellow; 2 – guide; 3 – base flange.

2.3 Linear-Motion Feedthrough

The linear-motion feedthroughs are not so perfect as the rotary-motion feedthroughs because of difficulty to create precise long-travel rigid guides in combination with ultrahigh vacuum sealing elements. Therefore, the number of different types of linear-motion feedthroughs is rather less than the number of different types of rotary-motion feedthroughs. The design of every type depends on the length of the travel, required precision, speed of the transference, etc.

The first solution which could occur to a designer is to use a cup-type Wilson-like seal like the one used in rotary-motion feedthroughs. However, this solution is

Fig. 2.4 The magnetic linear-rotary-motion feedthrough designed by Leisk Co. [17]: 1 – hermetic tube, made of non-magnetic material (body); 2 – moveable magnet (element of input movement); 3 – output stock.

unacceptable only for high vacuum linear feedthroughs because it is impossible to successfully retain the liquid penetration in the gap between cup-seal and linearly moving stock.

The traditional solution used in valves and other simple mechanisms with linear motion of the stock, consists in the sealing of the stock with a bellow. The length of traditional hydro-formed bellows can be extended only about 10–15% of its original length. The length of arc-welded bellows can be extended to 200% of its original length.

The linear-motion feedthrough based on the arc-welded bellow designed by Leisk Co. [16] is shown in Figure 2.3. This feedthrough is able to provide a precise motion on the distance of a hundred millimeters.

Another traditional solution for high vacuum technology is a two degrees of freedom magnetic feedthrough. The magnetic linear-rotary-motion feedthrough [17] designed by Leisk Co. is shown in Figure 2.4. This feedthrough uses a leak-tight non-magnetic tube (body) 1 which plays a role of linear guide for the movable magnet 2, which is magnetically coupled with the output shaft 3 inside the body. This feedthrough ensures the length of the travel close to the length of the non-magnetic tube minus the length of the carrier inside the tube. The precision and the rigidity of the feedthrough are limited by the parameters of magnetic fields of the used magnet. The disadvantage of this drive is that the outgassing high temperature treatment of the feedthrough must be done without the magnet.

A non-coaxial nut-screw couple based on untraditional design [18] can provide very long travel L ($>$ 1 meter) and overcomes the difficulty of the design. The scheme and general view of a non-coaxial planetary nut-screw linear-motion feedthrough is shown in Figure 2.5. The input rotary-motion of the leading eccentric bush 1 sets in circle motion without rotation of the leading nut 4 with zero pitch of thread. Nut 4 is coupled to the screw 5 being rotated. The nut's circle motion around the screw without the rotation of the nut makes the screw moving in straight lines.

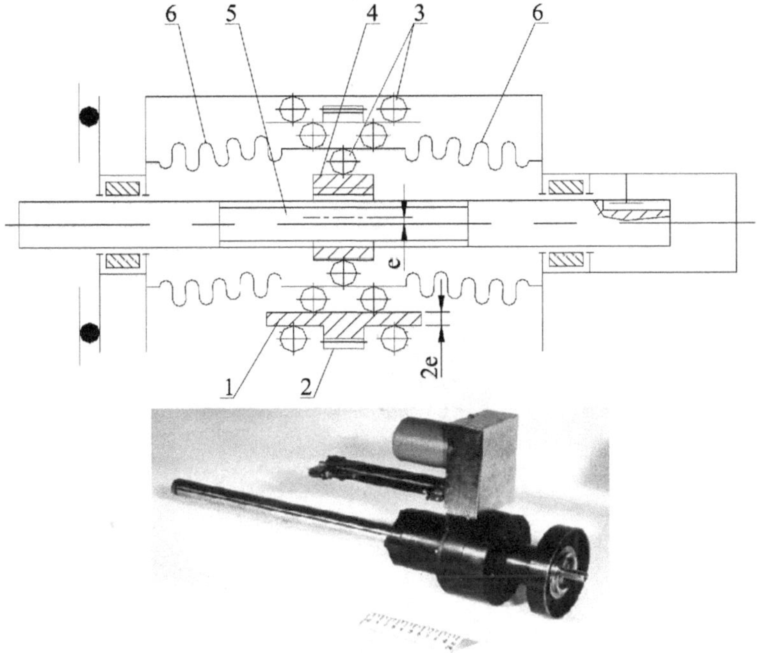

Fig. 2.5 (a) The scheme of non-coaxial planetary nut-screw linear motion feedthrough: 1 – leading eccentric bush; 2 – gear of the leading bush; 3 – ball bearings of the bush; 4 – leading nut with zero pitch of thread; 5 – output stock (screw); 6 – bellows for sealing. (b) The general view of the non-coaxial planetary nut-screw linear motion feedthrough.

The advantages of this kind of feedthrough are: small wear in vacuum because of small "friction way", high precision with linear error below 40 microns, high rigidity, the possibility to be heated up to 450°C without disassembling. The disadvantage is a possible small (5–20 microns) lateral motion of the output screw.

2.4 Manipulators

The manipulators with a large number of degrees of freedom usually are designed on the base of one-coordinate feedthroughs by means of their joining in series. The example of the base-mounting module is shown in Figure 2.6 [16]. The upper flange 2 of this module can be transferred along *X*, *Y*, *Z* coordinates in respect to the bottom flange 3. Both of these flanges are welded to the sealing bellow 1. The simplest two-coordinate magnet manipulator is shown in Figure 2.4.

Another example of two standard feedthrough modules assembled into one unit is the long-travel UHV manipulator designed by "MECA-2000" Co. [19] (Figure 2.7).

Fig. 2.6 The base-mounting module of multicoordinate manipulator designed by Leisk Co. [16]: 1 – sealing bellow; 2 – the upper mounting flange; 3 – the bottom mounting flange; 4, 5, 6 – the drives handles for x, y, z motions.

In this manipulator one of the bellows sealed rotary-motion feedthroughs (pos. 1) is used for linear motion of the stock 4. In this case a rack-and-gear drive is used. Another bellows sealed rotary-motion feedthrough 3 is used for rotation of the rod inside the stock.

The example of the seven coordinate bellows sealed manipulator designed by Vacuum Generators Company [6] is shown in Figure 2.8. The handles 1, 2, 3 are used for the carrier 4 transference along X, Y and Z axes. The handles 5, 6 are used for inclinations of the stock 7. The handle 9 is used for linear movement of rod 10.

The example of two-coordinate long-travel transporting manipulator of UHV technological module without friction couples in vacuum [20] is shown in Figure 2.9. The carrier 1 situated in the atmosphere ensures the long-travel linear motion of the clam-shell, which is situated inside the vacuum module, using the screw drive 2. The rotation drive (harmonic rotary feedthrough) 3 is used for the inclination of the transporting clam-shell .

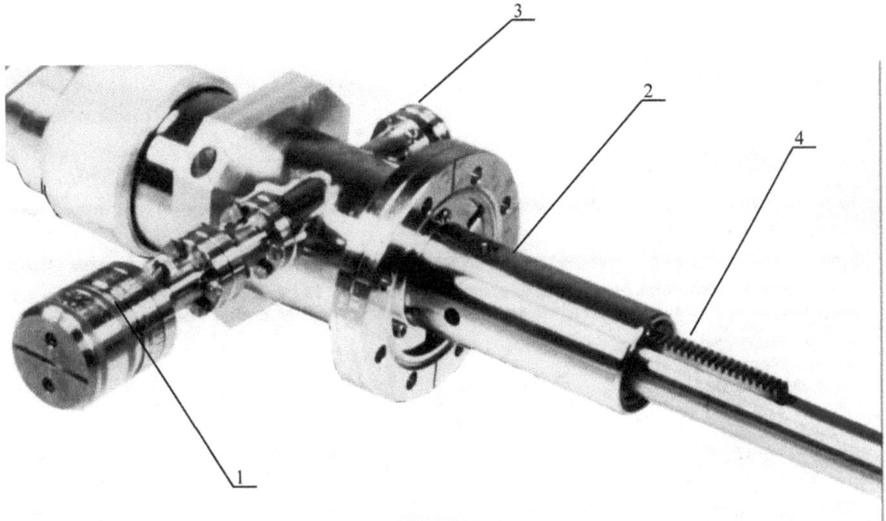

Fig. 2.7 The two-coordinate manipulator of "MECA-2000" Co. [19]: 1 – bellow sealed rotary-motion feedthrough, which is used for the rack (stock) linear transference; 2 – cantilever; 3 – bellow sealed rotary-motion feedthrough, which is used for the rod rotation inside the stock; 4 – rack.

The structural materials for design of ultrahigh vacuum feedthroughs and manipulators are limited to traditional materials for UHV technology (stainless steel, Kovar, glass, vacuum copper, vacuum ceramic, etc.).

2.5 Micro Mechanisms

The new and promising direction in which the design of precise vacuum mechanisms is moving can be illustrated by a design of autonomous micro-electro mechanical systems (MEMS) [21–23] built on the base of finished silicon monocrystal, polycrystalline sample or photopolymerization of compounds. These small mechanisms consist of several components and units (rotors, gears, chains, pumps, micro motors, etc.) all of which are about a few microns in size. They are able to function successfully both in ultrahigh vacuum and in normal atmospheric conditions because of their small outgassing rate and high allowable outgassing temperature 200–400°C. The friction couples of these mechanisms work in conditions of so called "dry friction". The friction force determines the precision and reliability of MEMS. Figure 2.10 shows the micromachine created by Sandia National Laboratories [21–23]. Figure 2.10a shows the general view of the silicone micromachine. Its overall dimensions are below 0.3 mm. Figure 1.10b shows silicone components of the drive at a larger scale.

Fig. 2.8 The seven degrees of freedom bellow sealed manipulator designed by Vacuum Generators Co. [6]: 1, 2, 3 – the handles for the carrier transference along X, Y, Z axes; 4 – carrier; 5, 6 – the handles for the stock inclination; 7 – stock; 8 – handle for the stock rotation; 9 – handle for the rod (inside the stock) linear transference; 10 – rod of the stock.

Design of scanning mechanisms for scanning microscopy and MEMS is based mainly on the use of a piezo ceramic, monocrystal and polycrystalline silicon.

The short review of vacuum installation kinematic and used mechanisms shows that a great number of mechanisms can be reduced to a smaller number of rotary-

Fig. 2.9 The two-coordinate long-travel transporting manipulator based on UHV technological module [20]: 1 – carrier; 2 – the screw drive of linear transporting movement of the clam-shell; 3 – the rotation drive of the transporting clam-shell inclination.

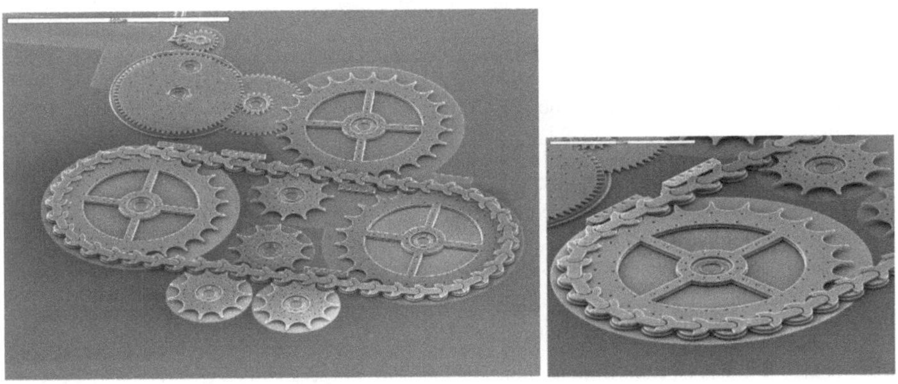

a) b)

Fig. 2.10 Micro electro mechanical system (MEMS), created by Sandia National Laboratories (USA): (a) silicone micromachine; (b) silicone details of the drive.

and linear-motion feedthroughs. Kinematic elements can be limited to the following types of kinematic pairs:

- friction couple "bush-shaft";
- rigid and flexible gears;
- coaxial and non-coaxial "nut-screw" couples;
- ball-bearings;

- chain-drives.

The work of all these kinematic pairs is based on a friction process. The designer's task is to minimize the outgassing flow from these pairs and to minimize the negative influence of a vacuum on the reliability of these mechanisms.

In the next chapter we will consider the nature of friction forces in vacuum and the nature of the gas exchange processes which determine the reliability of vacuum mechanisms in general.

References

1. A. Roth, *Vacuum Technology*, 3rd ed., North-Holland, Amsterdam, 1990.
2. Booklet of Edvards Vacuum Equipment Co., Crawly (England), V. 12, #91, p. 132.
3. R.R. Wilson, A vacuum-tight sliding seal, *Rev. Sci. Instr.* **12**, #91, 1941, pp. 9193.
4. Booklet of ULVAC, Rotary motion feedthrough, Model TDU-8, NE2302, Tokyo, 1977, 2 pp.
5. Booklet of Leisk, RM Series Rotary motions, 04.003.180, Burgess Hill, England,1981, 2 pp.
6. Booklet of Vacuum Generators, "Vacuum Component Booklet. Rotary Drives", Hastings (England), 1981, 320 pp.
7. Booklet of Varian, "Varian Vacuum Products", Italy, 1995/1996, pp. 445–446.
8. Booklet of Riber, UHV Short Form, 608.10C.22, Manipulation: Single Motion Feedthrough, RUEL Malmaison (France), 1988, 38 pp.
9. Booklet of Riber.
10. Booklet of Leybold-Heraeus, Schiebedruckfuhring. Kat.# 882/29, Hanau (Germany), 1986, pp. 11–12.
11. Booklet of MECA, Traversees de Rotation Brochure #0913, Vermouvillet (France), 1990, 2 pp.
12. E.A. Deulin, N.S. Vagin, Wear and longevity of UHV harmonic rotary feedthrough, ECASIA-97, Goteborg (Sweden), 1997, pp. 1147–1150.
13. E.A. Deulin, N.S. Vagin, UHV harmonic rotary feedthrough of high reliability, *Vacuum Technique and Technology* **8**(12), 1997, pp. 17–23 [in Russian].
14. E. De Haas, The harmonic drive as UHV rotary motion feedthrough, *Nuclear Instr. and Methods* **137**, 1976, pp. 435–439.
15. Y. Hayashi, Development of the harmonic drive rotary feedthrough, *Vacuum* **41**, 1990, pp. 1948–1950.
16. Booklet of Leisk, Custom Built Motions, 04.006.984, Burgess Hill (England), 1981, 2 pp.
17. Booklet of Leisk, MIRM Series Magnetically Coupled Linear-Rotary Motion, 04.005.287, Burgess Hill (England), 1981, 2 pp.
18. D.V. Bushenin, E.A. Deulin, S.P. Nosatov et al., *Design of Screw Mechanism. Methodological Recommendations*, Vladimir, 1979, 129 pp.
19. Booklet of "MECA", Transfer Rods. Preliminary brochure 4015, Vermovillet (France), 1990, 2 pp.
20. E.A. Deulin, *Mechanics and Physics of Precise Vacuum Mechanisms*, Vladimir State Univ., Vladimir, 2001, Vol. 1, 176 pp.
21. J. Brinker, Mega benefits, microsystems, *Sandia Technology* **1**(2), 1999, pp. 2–6.
22. S. Rodgers, J. Sniegowski, Building a better microsystem, *Sandia Technology* **1**(4), 1999, pp. 8–9.
23. J. Vitko, At Sandia. Microsystem abound, *Sandia Technology* **1**(2), 1999, pp. 8–16.

Chapter 3
Friction in Vacuum

3.1 Friction Coefficients of Different Materials in Atmosphere and in Vacuum

According to the most famous publications [1–7] in the field of vacuum mechanics, the vacuum is considered as an aggressive medium, which increases friction forces and accelerates the wear of friction couples of vacuum mechanisms. For the illustration of this position the known [3, 5, 6, 8] data about friction coefficients in atmosphere and in vacuum for traditional materials are shown in Table 3.1.

We can see that the previous publications does not show the difference of friction coefficient values in the low, middle, high and ultrahigh vacuum.

As a matter of fact, both vacuum degree and temperature also as friction speed, contacting load, have a great influence on the behaviour of a friction coefficient. In this chapter it will be shown that the physical nature of the friction coefficient varies in vacuum.

3.2 Dry Friction Laws in Atmosphere and in Vacuum

The friction laws, which show the independence of the friction coefficient from the speed and load, were first discovered by Leonardo da Vinci in 1508, and this discovery was repeated by G. Amontons in 1699.

It was shown that the area of contact A_c is a result of the micro roughness deformation at yield stress σ_t:

$$A_c = N/\sigma_t, \tag{3.1}$$

where N is the normal load.

In this case, the force, which can ensure the microroughness cutting in area of the contact (i.e. friction force) is:

$$F = A_c \times \tau_c, \tag{3.2}$$

E.A. Deulin et al., *Mechanics and Physics of Precise Vacuum Mechanisms*,
Fluid Mechanics and Its Applications 91, DOI 10.1007/978-90-481-2520-3_3,

Table 3.1 Friction coefficients of the metals, solid lubricants and antifriction materials.

#	Friction materials	Friction coefficient in air	Friction coefficient in vacuum
1	Fe-Fe	0.3	1.9
2	Fe-Mg	1.0	0.6
3	Fe-Cd	1.5	0.4
4	Fe-Pb	0.9	0.4
5	Stainless steel- Stainless steel	0.5	2.9
6	Stainless steel-Cu	-	0.3
7	Stainless steel-Al	0.4	0.3
8	Stainless steel- kovar	-	0.4
9	Stainless steel- brass	0.4	0.8
10	Stainless steel-Mo	-	0.8
11	Stainless steel- lanthanum	-	0.8
12	Stainless steel-Ni	-	0.8
13	Stainless steel-"Dimolit-1" (antifriction film)	-	0.05
14	Stainless steel- Teflon	-	0.2
15	Stainless steel-Cu+MoS_2 (antifriction film)	-	0.2
16	Stainless steel- Stainless steel+MoS_2 (antifriction film)	-	0.2
17	Stainless steel-Si (polished)	-	0.2
18	Stainless steel-Ge (polished)	-	0.2
19	Stainless steel- ground ceramic "22XC"	-	0.3
20	Stainless steel- polished ceramic "22XC"	-	0.4
21	Stainless steel- polished glass ceramic "CT-50-1"	-	0.5
22	Stainless steel- ground glass ceramic "CT-50-1"	-	0.3
22	Stainless steel- polished glass "C48-3"	-	0.5
23	Stainless steel- ground glass "C48-3"	-	0.3
24	Chromium steel - chromium steel	0.5	0.5
25	Chromium steel - Cu	0.7	1.0
26	Cu-Cu	0.5-1.0	4.8-21.0
27	Cu-Ni	0.6	1.5-2.0
28	Cu-Ta	0.4	0.4
29	Cu- W	0.3	0.6
30	Al-Al	0.8	1.6-2.2
31	Al-Ni	-	2.4
32	Al-Cu	-	1.5
33	Al- Ag	-	2.2
34	Brass-brass	0.4	0.7
35	Beryllium bronze-beryllium bronze	0.7	1.1
36	Beryllium bronze- brass	0.4	0.9
37	Ni-Ni	-	4.9
38	Ta-Ta	0.2	4.7
39	W-W	0.2	2.5
40	Ag-Ag	-	3.9
41	Co-Co	-	0.3-0.5
42	Cr-Cr	0.6	3.0
43	Mo-Mo	1.0	2.5
44	Au-Au	0.6	4.5
45	Zn-Zn	1.0	3.0
46	Sn-Sn	1.0	1.0
47	Ti-Ti	-	4.2
48	Pt-Pt	-	3.5
49	Nb-Nb	-	4.2
50	Zr-Zr	-	1.5
51	Chromium steel – MoS_2 film (vacuum deposition)	-	0.06
52	Chromium steel – MoS_2 film (friction deposition)	-	0.06
54	Chromium steel – MoS_2 film +epoxy compound	-	0.1
55	Chromium steel – MoS_2 film + organosilicon compound	-	0.2
56	Chromium steel – MoS_2 film + sodium silicate	-	0.1
57	Chromium steel – MoS_2 film +TSO_2	-	0.2
58	Cu-MoS_2	0.2	0.07
59	Cu-CdS_2	0.5	0.2
60	Cu-ViJ	0.4	0.3-0.5
61	Cu- phthalocyanine	0.4	0.4
62	Chromium steel-MoSe	0.2	-
63	Chromium steel-WSe_2	0.2	0.1
64	Chromium steel-TaSe	0.2	0.1

where τ_c is the cutting breaking point, MPa.

Summarizing Equations (3.1) and (3.2) the friction coefficient is:

$$f = \frac{F}{N} = \frac{N \cdot \tau_c}{N \cdot \sigma_\tau} = \frac{\tau_c}{\sigma_\tau}. \tag{3.3}$$

These formulae show that the friction coefficient does not depend on the area of contact, but depends only on strength parameters of contacting materials:

$$f = \frac{\tau_c}{\sigma_\tau} + \beta \tag{3.4}$$

3.3 The Main Factors, which Determine the Surface Coverage at "Dry" Friction

The previous creators of modern theory of friction [1–6] make the explanation of dry friction mechanisms with help of a so-called "third body". They consider the different components in the role of a "third body": oxides, traces of lubricants or organic materials, thin layers of the remnants of destroyed or dislocated surfaces, and they forget about the (in our opinion) main component -the film of the adsorbed gases.

According to our concept the main component of the "third body" is the film of adsorbed gases, mainly water, which is always present on the surface. According to Langmour's experiments [9] the water coverage on a glass may reach 50–55 monolayers, which are equal to 15–16 nm.

The creators of the friction theory call the tangential forces, which link the layers as the "molecular component" of the friction force. We consider that the nature of this force is more complex and it includes at macro range a capillary and an adhesive component [10, 11], also as at the micro range: a kinetic, a viscous, an electrostatic, a correlation and a "change" component [12, 13].

We consider the film of adsorbed gases strong enough to keep the contacting load and to prevent direct (cohesive) contacting of surfaces. Thus, the adsorption energy (for the first layer) of "water-silicon" $E_S = 1.65$ J·m^{-2} [12] and it is only a half of the linking energy "Si–Si" in silicon monocrystal, which is $E_S = 3.24$ J·m^{-2}. Let us remember that ultimate stress for silicon monocrystal is $\sigma_B = 1200$ MPa, that is why ultimate stress for the first adsorbed water monolayer may be expected close to $\sigma_B = 600$ MPa. It means that the first adsorbed monolayer's ultimate stress is close to stainless steel, type 304, for which $\sigma_B = 630$ MPa.

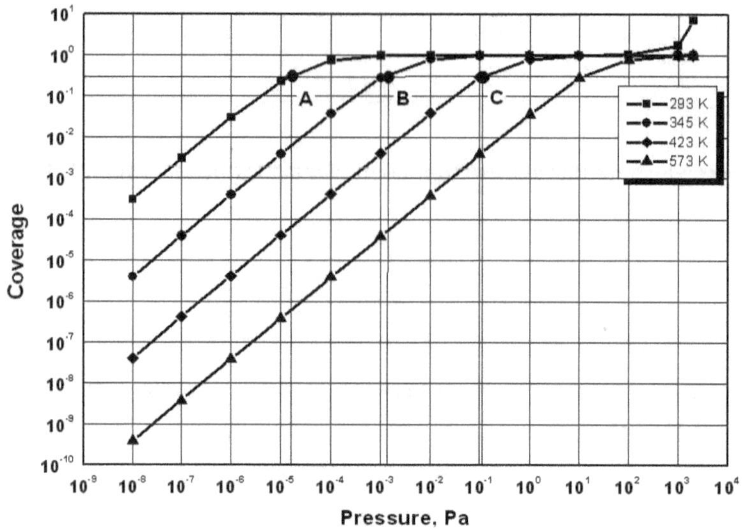

Fig. 3.1 The isotherms of the "water-silicon" system, calculated according BET equation.

3.3.1 Influence of the Residual Pressure and Temperature

The adsorbed gas film thickness (coverage Θ parameter) can be calculated with adsorption isotherm equations. The Henry and Langmour isotherms [8, 14] are used only for small pressures, the Dubinin–Radushkevitch one [15] can be used only for porous, but not for sliding contact. In the case of the polymolecular adsorption the surface coverage Θ may be calculated with the help of the BET [16] equation:

$$\Theta = \frac{\frac{P}{P_S} \cdot \exp\left[\frac{E_S - E_L}{R \cdot T}\right]}{\left(1 - \frac{P}{P_S}\right)\left[1 + \left(\left(\frac{E_S - E_L}{R \cdot T}\right) - 1\right) \cdot \frac{P}{P_S}\right]}, \tag{3.5}$$

where P is the residual pressure of the adsorbed gas, Pa; P_s is the saturated pressure of the adsorbed gas, Pa; E_S is the adsorption energy, J·mol^{-1}; E_L is the energy of evaporation for the adsorbed gas, J·mol^{-1}; R is the gases constant, $R = 8.34 \cdot 10^3$ J·mol^{-1}; T is the surface temperature, K.

Figure 3.1a shows a lot of isotherms for the water, adsorbed on silicon, according to Equation (3.5). Thus, if we consider that the "dry friction" behaviour depends mainly on coverage of Θ parameter, the temperature variation has to cause the friction variation, which is similar to the friction variation caused with the residual pressure variation, as it follows from Figure 3.1.

It is known that the BET equation is incorrect in the region close to a saturated pressure. In our case at the room temperature (293 K) and 50% RH (water vapour partial pressure $1.17 \cdot 10^3$ Pa) original BET equation gives $\Theta = 1.7$ for water adsorbed on Si, at the same time the publications describing Atomic Force Microscopy (AFM) results [17] show that the cumulative coverage thickness is 30 Nm (i.e. cumulative coverage value is about $\Theta_\Sigma \approx 100$, coverage value is $\Theta_\Sigma \approx 50$). In correspondence with these figures the BET equation was corrected by the authors and presented in the next form:

$$\Theta = \frac{\frac{P}{P_S} \cdot \exp\left[\frac{E_S - E_L}{R \cdot T}\right]}{\left(1 - \frac{P}{P_S}\right)\left[1 + \left(\exp\left(\frac{E_S - E_L}{R \cdot T}\right) - 1\right) \cdot \frac{P}{P_S}\right]} \cdot k\ , \tag{3.5a}$$

where k is the coefficient, which takes into account the water dissociation processes on the surface and the action of OH radicals.

It was found that the best equation (in accordance with the experimental results) is:

$$P = a + b \cdot p^2,$$

where $a = 1$ and $b = 2.75 \cdot 10^{-5}$.

It should be noted that the surface temperature determination at friction in the upper surface monolayer is a difficult task, because the experiments allow only very average temperature means. The theoretical calculations [18–20] give different results. For example, the calculation of "flash temperature" according to [1] for friction couple "steel-steel" at next conditions: sliding speed $V = 0.013$ m·s^{-1}, contacting pressure $\sigma_c = 617$ MPa, heat conductivity coefficient $k = 14.5$ Wt·m^{-1}·K^{-1}, friction speed $V = 3.4 \cdot 10^{-2}$ m·s^{-1}, and friction coefficient $f = 0.3$ gives $T_f = 311.6$ K. On the other hand, the calculation according to [19] gives $T_f = 410$ K (at average temperature $T_a = 310$ K).

3.3.2 Influence of the Sliding Velocity and Roughness Geometry

First of all, the friction parameters, first of all friction speed, have a great influence on surface temperature. Besides, a distance between the micro roughness together with the friction velocity determine the period τ_c between the consequence contacting actions:

$$\tau_c = \frac{L}{V}, \tag{3.6}$$

where: L is the average distance between micro roughness, m; V is the friction speed, m·s^{-1}.

At this very short period the processes of gas exchange takes place on the surfaces, and the coverage varies very quickly. The surface, heated to "flash temperature" T_f after contacting is finished, becomes cleaner [9], because of "flash" gas release at time of deadsorption/release of gas:

$$\tau_d = \tau_0 \cdot \exp\left(\frac{E_S}{R \cdot T_f}\right), \tag{3.7}$$

where: τ_0 is the coefficient, corresponding to molecular oscillation, $\tau_0 \approx 10^{-13}$, s, E_S is the release (adsorption) energy; T_f is the "flash temperature" at contacting, K.

We will indicate the result of gas release from the surface as Θ_f and that corresponds to the temperature T_f according to Equation (3.5). The numbers of the molecules, which can be adsorbed in addition to coverage Θ_f at average temperature T_a at period $(\tau_c - \tau_d)$ are:

$$N_S' = \frac{P \cdot V_a \cdot \alpha}{4k \cdot T_a}(\tau_c - \tau_d), \tag{3.8}$$

where: T_a is the average temperature of the micro roughness before new contacting action, K; V_a is the average velocity of the gas molecules, m·s^{-1}; α is the sticking coefficient; k is the Boltzmann coefficient, $k = 1.38 \cdot 10^{-23}$ J·K^{-1}.

The coverage of the surface (at the top of the micro roughness) before a new contacting action may be expressed:

$$\Theta = \frac{N_S}{N_{1\Pi}} + \Theta_f = \frac{P \cdot V_a \cdot \alpha}{N_{1\Pi}} + \Theta_f = \frac{P \cdot V_a \cdot \alpha}{N_{1\Pi} \cdot 4 \cdot k \cdot T_a} \times \left(\frac{L}{V} - \tau_0 \cdot \exp\left(\frac{E_S}{RT_f}\right)\right) + \Theta_f, \tag{3.9}$$

where $N_{1\Pi}$ is the number of the "vacant places" for the molecules, being adsorbed, m^{-2}.

After a small transformation, Equation (3.9) may be written as

$$\Theta \approx \frac{P \cdot d_0^2 \cdot \alpha}{2.82 \cdot \sqrt{m \cdot k \cdot T_a}} \left(\frac{L}{V} - \tau_0 \cdot \exp\left(\frac{E_S}{RT_f}\right)\right) + \Theta_f, \tag{3.9a}$$

where m is the mass of the adsorbed molecule, kg; d_0 is the diameter of the adsorbed molecule, ($d_0 \approx 3 \cdot 10^{-10}$ m for water)

Thus, we can see that the coverage depends on the external parameters of the used materials (P, α, T_a) and on the friction couple parameters (V, L, E_S). Equations (3.9), (3.9a) help to estimate the amount of gas, which can be adsorbed again at T_a temperature after the gas release process at T_f temperature.

Let us consider the friction force and friction coefficient variation as a function of cumulative decrease of coverage.

The theoretical calculations were done for the surface profile of the polished silicon monocrystal, which is shown in Figure 3.2, and then the theoretical results were compared with the experimental result [21] for the same sample.

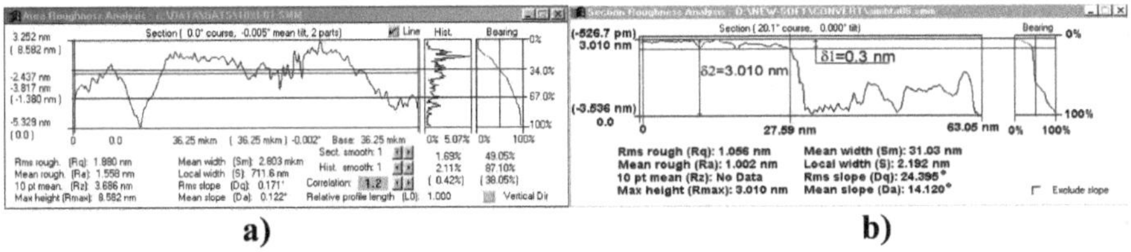

Fig. 3.2 One-dimensional cross-section of Si surface (plane 100), made with STM [20]: (a) scan which was used for theoretical calculations (scan dimensions 36.25 × 36.25 mic, maximum micro roughness high 8.5 nm); (b) scan which shows the ideally smooth top of micro roughness (deviation is less than 0.3 nm).

3.4 The Theoretical Analysis of Friction in the Different Ranges of Coverage

Dynamic friction coefficient analysis was based on the created physical model with use of the real profile of the polished Si monocrystal. The example of the searched surface profile [20], which was achieved with STM [22], is shown in Figure 3.2. It is a surface (in plain 100) of a polished Si monocrystal covered with SiO_2 film.

On the right-hand side of the presented profiles (Figure 3.2) we can see the variation of contacting probability $P(x)$ of this surface with an ideal one, as a function of depth ("bearing" parameter).

At the first stage of the experiments the sliding speed was about $V \approx 0.01$ m·s^{-1}, "flash temperature" was close to room temperature ($T_f - T_a \approx 3$°C) and its influence on the friction process was considered to be negligible.

"Mechanical component" of friction coefficient f_M was established accordingly [1]:

$$f_M \approx \sqrt{\frac{h}{r}}, \tag{3.10}$$

where h is the depth of mutual penetration of the micro roughness, m; r is the equivalent radius of the micro roughness, m.

This component was considered as negligible, because its value is less than 0.01% of the friction coefficient for the profiles, considered in Figure 3.2.

In the theoretical model the assumption was made that if the adsorbed layer thickness is equal to $\Theta \cdot d_0$ (where Θ is the coverage, d_0 is the molecule diameter) the length of contacting zone in one dimension is equal to the length of the accumulated cross-section of the micro roughness at the depth $h = \Theta d_0$ (from the top of the highest micro summit).

If the area of the considered surface makes contact with an ideal one, the depth h may be expressed as $A \cdot P(x) \cdot P(y)$, where A is the "geometry" (nominal) contacting surface; $P(x)$, $P(y)$ is the probability of contacting along the X and Y axes.

If we take into account Langmour's result [9] which shows that the number of adsorbed water monolayers (coverage Θ) on the glass in the atmosphere may reach $\Theta = 50$–55, we can see that the adsorbed layer thickness on a surface can reach

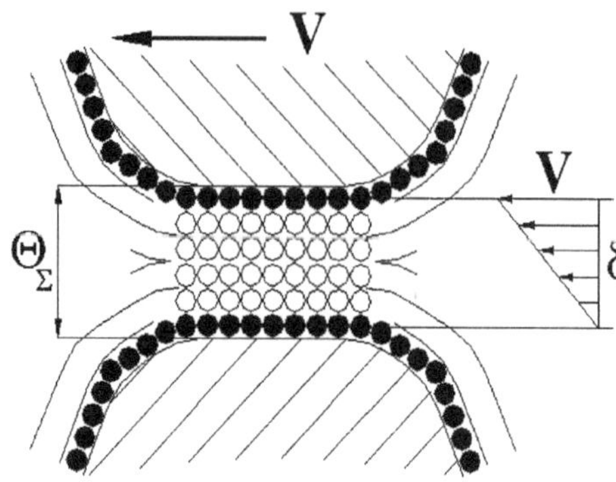

Fig. 3.3 Scheme of the viscous friction.

$50 \cdot d_0 \approx 15$ nm (where d_0 is a diameter of adsorbed gas molecules $d_0 \approx 0.3$ nm). In this case contacting probability $P(x)$ is close to 1. In vacuum, or at high temperature, coverage decreases (for example till $\Theta \approx 0.3$ at residual pressure $P = 1$ Pa).

Let us consider dynamic friction force variation step by step, according to decrease of coverage.

3.4.1 Viscous Component of a Friction Force

As it was shown before, the viscous component (also known as capillary component) exists at total coverage $\Theta_M > 2$.

Viscous component of friction force can be expressed with the help of Newton's equation:

$$F_V = A \cdot P^2(x) \cdot \mu \cdot \frac{dV}{dz} = A \cdot P^2(x) \cdot \mu \cdot \frac{V}{\delta}, \tag{3.11}$$

where A is the geometry (nominal) contacting area, m^2; $P_1(x)$, $P_1(y)$, $P_2(x)$, $P_2(y)$ is the contacting probability of the first and second surfaces along axis X and Y, correspondingly; z is the depth coordinate, m; μ is the coefficient of dynamic viscosity of adsorbed gas, kg·s^{-2}·s^{-2}·m^{-1}; V is the sliding velocity, m·s^{-1}; Δ is the accumulated thickness of adsorbed gas, m.

The first layer of the adsorbed gas, fixed on the surface with energy E_S is considered in quasi-solid state [16, 23]. That is why it cannot be taken into account in Equation (3.11), as it is shown in Figure 3.3.

In this case the accumulated thickness of quasi-liquid adsorbed gas is

$$\delta = \Theta_\Sigma - 2 \cdot d_0, \tag{3.12}$$

where Θ_Σ is the accumulated coverage, d_0 is diameter of adsorbed molecules.

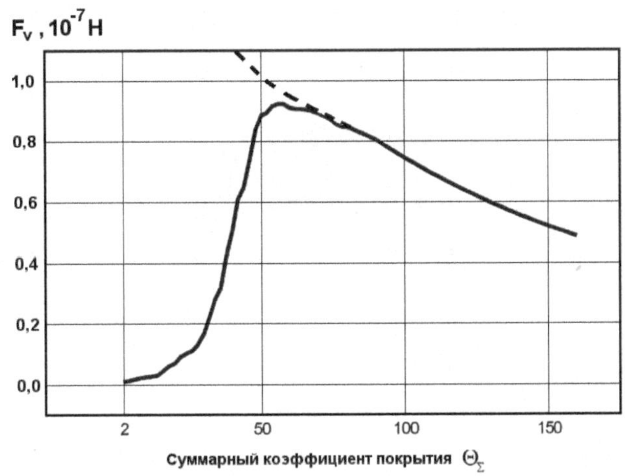

Fig. 3.4 Viscous component of friction force F_V as a function of Θ, for the measured sample (see Figure 3.2).

The results of F_V calculation as a function of a total coverage Θ_Σ are presented in Figure 3.4. Parameters of friction were: sliding speed $V = 0.01$ m·s^{-1} dynamic viscosity of adsorbed gas (water, because it is adsorbed with the highest energy), $\mu = 0.001$ kg·s^{-1}·m^{-1}; the highest value of micro roughness $h_{\max} = 8.5$ nm; the maximum thickness of quasi-liquid adsorbed gas $\delta_{\max} = (\Theta_{\Sigma\max} - 2) \cdot d_0 = 45$ nm, where $\Theta_{\Sigma\max} = 150$.

3.4.2 Capillary Component of a Friction Force

The capillary component of a friction force F_C is the accumulated result of all capillary forces action around all contacting points (spots) as it shown in Figure 3.5. Capillary forces act on the boundaries of contacting spots and the force vector direction depends on watering angle φ also as on inclination angle α of the micro roughness profile:

$$F_C = \sigma \cdot \sum_i l_i = l_\Sigma \cdot \sigma, \tag{3.13}$$

where l_i is the length of a shore line along the boundaries of contacting spot, m; i is the number of contacting spots; σ is the surface tension, N·m^{-1}; l_Σ is the accumulated length (coastline) of the boundary line of the micro roughness, calculated for the given depth $z = \delta/2$ of the analyzed profile of the contacting surface, m.

The variation of capillary force F_C as a function of a total coverage for the sample is shown in Figure 3.6.

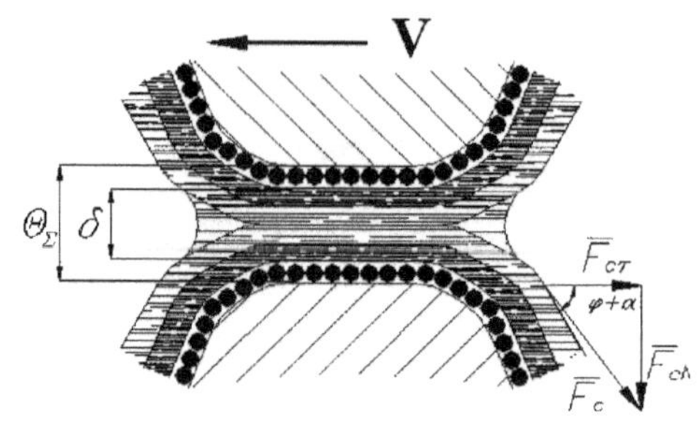

Fig. 3.5 Scheme of capillary forces action.

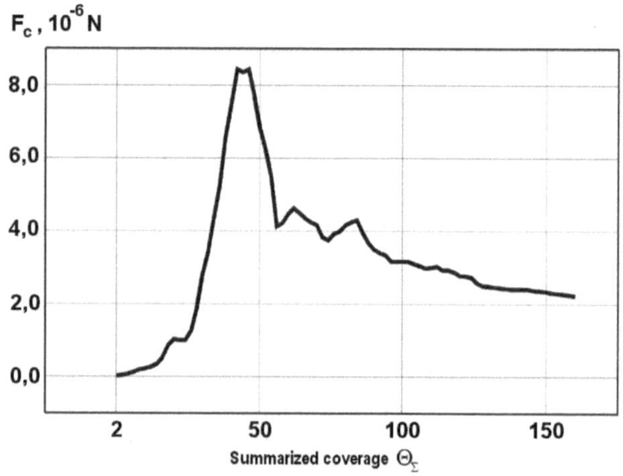

Fig. 3.6 Capillary force F_C as a function of total coverage Θ_Σ for the measured sample (see Figure 3.2).

The tangential component of the capillary force $F_{C\tau}$ depends not only on φ and α angles but also on the angle ψ between the force F_C and sliding velocity V vectors:

$$F_{c\tau} = \sigma \cdot \cos(\varphi + \alpha) \cdot \sum_i l_i \cdot \int_{\psi=0}^{2\pi} \cos\psi \cdot d\psi \,. \tag{3.14}$$

For the contacting spots with a symmetrical shape in respect to vector V, the expression (3.14) can be written as

$$F_{c\tau} = 0.5 \cdot \sigma \cdot \cos(\varphi + \alpha) \cdot l_\Sigma \,. \tag{3.15}$$

The normal component of the capillary force F_{Cn} can be expressed:

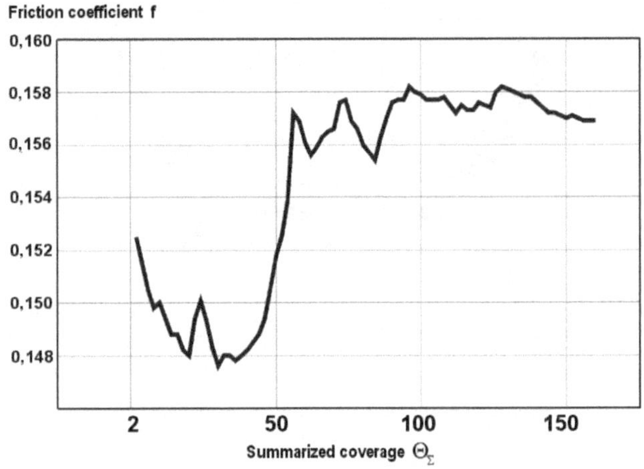

Fig. 3.7 The results of friction coefficient F_{CV} calculation as a function of summarized coverage Θ_Σ for the measured sample (see Figure 3.2).

$$F_{Cn} = l_\Sigma \cdot \sigma \cdot \sin(\varphi + \alpha). \tag{3.16}$$

The capillary force normal component F_{Cn} together with normal load N defines the friction coefficient.

The dynamic friction coefficient f_{CV} calculated in this range may be expressed:

$$f_{cv} = \frac{F_{C\tau} + F_v}{N + F_{cn}}, \tag{3.17}$$

where N is "external" normal load at friction, N.

The normal "external" load N is the result of the sample weight action. The results of the calculation of friction coefficient F_{CV} as a function of accumulated coverage Θ_Σ are presented in Figure 3.7. The parameters of friction were: : surface tension of adsorbed gas (water) $\sigma = 72.8$ H·M^{-1}, watering angle $\varphi = 60°$, inclination angle of the micro roughness $\alpha \approx 0.636°$.

3.4.3 Adhesive-Viscous Friction

When the accumulated coverage decreases and becomes $\Theta_\Sigma = 2$ the viscous component of friction dV/dz reaches its maximum value $dV/dz = V/d_0$, where d_0 is a distance between the centers of the molecules of the first adsorbed monolayers, which are considered as a solid (Figure 3.8).

In the case of adhesive-viscous friction we have come to view the mechanics of friction on the scale of one atom and the laws of classical mechanics become unapplicable. For the prolongation of theoretical analysis we may use the energy

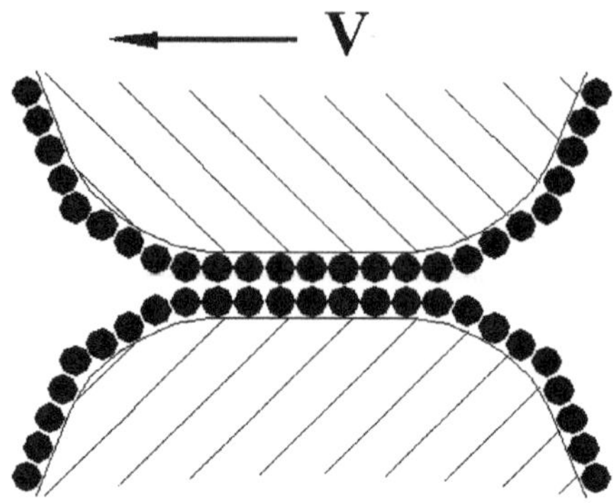

Fig. 3.8 Scheme of adhesive-viscous friction.

theory [24]. The friction force and the friction coefficient can be calculated using the adapted energy theory.

In this case an attractive energy E_L (evaporation energy) between two layers of adsorbed gases on the contacting surfaces decreases according to $(\Theta_\Sigma - 1)$ decreasing, and it leads to the corresponding friction force decreasing. According to this theory, the very contacting action is accompanied by the generation of heat energy, which is equal to the total adsorption energy of the adsorbed gases molecules, which are squeezed in the contacting slot and are situated in the contacting plane. In case of adhesive-viscous friction, when, $1 < \Theta_\Sigma \leq 2$, Figure 3.8, friction plane is situated between the two first monolayers of the adsorbed gases. That is why the attractive energy (adsorption energy) in this friction plane is equal to the evaporation heat E_L (according to BET equation [16]) plus melting energy, because the first monolayers are considered as solids.

Let us consider that friction occurs as a result of mutual movement of contacting surfaces, consisting of regular contacting spots, (micro roughness with flat upper surfaces), which are situated on the surface with step L_X, L_Y and have the contacting length l_X, l_Y in X and in Y directions Figure 3.9.

In case that the upper surface is transferred at distance X, the contacting actions number n_X for a relatively smooth micro roughness is

$$n_X = \frac{X}{L_X}, \tag{3.18}$$

$$A_r = A \cdot P_1(X) \cdot P_1(Y) \cdot P_2(X) \cdot P_2(Y). \tag{3.19}$$

In case of friction of two isotropic surfaces with the same geometry of the micro roughness:

$$A_r \approx A \cdot P^4(X). \tag{3.20}$$

The heat energy, which is generated after every single contacting action of these surfaces:

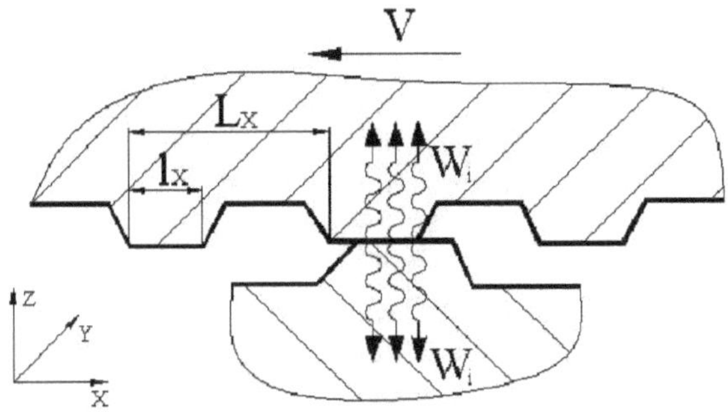

Fig. 3.9 Scheme for explanation of energy theory of friction.

$$W_1 = (E'_L + E'_M) \cdot A \cdot P^4(X), \tag{3.21}$$

where W_1 is the heat energy generated at one contacting action; E'_L is specific evaporation energy (of adsorbed gases), J·m^{-2}; E'_M is specific melting energy (of adsorbed gases), J·m^{-2}.

Contacting process is an exothermic process. The generated heat goes out of the contacting zone by means of heat conductivity. The heat energy, generated on friction surfaces during movement over distance X is:

$$W_X = n_X \cdot (W_1 + W_2), \tag{3.22}$$

where $W_i = W_1, W_2$ is adsorption energy, generated on the contacting spots of the first and second surfaces, J.

The process of the micro roughness tops removing from the contacting spots is the endothermic process and the system spends the mechanical equivalent of the energy described by Equations (3.21) and (3.22) for the movement continuation. For the continuation of the movement it is necessary to apply the friction force F_{AV}, which for the movement over the X distance for the similar isotropic surfaces is equal:

$$\begin{aligned} F_{AV} &= \frac{W_X}{X} = \frac{(E'_L + E'_M) \cdot A \cdot P^2(X) \cdot P^2(Y) \cdot (\Theta - 1)}{L_X} \\ &\approx \frac{(E'_L + E'_M) \cdot A \cdot P^4(X) \cdot (\Theta - 1)}{L}, \end{aligned} \tag{3.23}$$

where L is the mean step size of the micro roughness, m; $P(X)$ is the probability of the surface cross-section coincidence of the considered surface with an ideal one.

We can see from Equation (3.23) that for isotropic surfaces the friction force is different for motion in different directions. This position is in accordance with experiments of other authors [25, 26]. We can see also that the friction force F_{AV} depends on parameter L, which can be considered as a fractal parameter [27, 28].

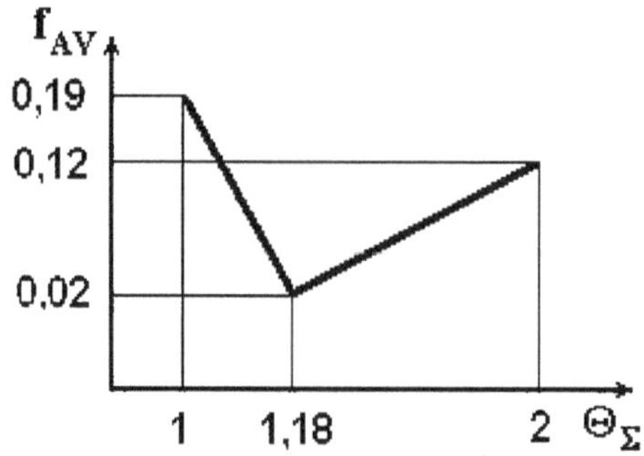

Fig. 3.10 Coefficient of adhesive-viscous friction F_V as a function of Θ_Σ, for the measured sample (see Figure 3.2).

Friction coefficient for the considered sample of weight N (taking into account the capillary component $F_{\rm CN}$) can be expressed as

$$f'_{\rm AV} = \frac{F_{\rm AV}}{N + F_{\rm CN}}. \tag{3.24}$$

According to Equation (3.24) the friction coefficient $f_{\rm AV}$ linearly depends on the contacting energy W_X that is proportional to $(\Theta_\Sigma - 1)$ quota.

After Θ_Σ decreases to critical value $\Theta_{\Sigma\rm AV}$

$$\Theta_{\Sigma\rm CAV} = \frac{N + F_{\rm CN}}{\sigma_B \cdot A \cdot P^4(X)} + 1, \tag{3.25}$$

where σ_B is breaking point of the first monolayer of the adsorbed gases, $\sigma_B =$ 600 MPa for water on Si.

For the sample analyzed (see Figure 3.2), we have: $N = 1,2 \cdot 10^{-8}$N, $A = 1.3 \cdot 10^{-9}\ \rm m^2$, $P^4(X) = 5.06 \cdot 10^{-8}$, $F_{\rm CN} = 3.38 \cdot 10^{-10}$N, $\sigma_B = 600$ MPa. After calculation we receive the critical coverage $\Theta_{\Sigma\rm CAV} = 1.302$.

The solid structure of the first (solid) monolayer of the adsorbed gases becomes destroyed (on one of the surfaces), because the normal pressure in the contacting zone exceeds the loading possibility of a sorbate on $(\Theta_\Sigma - 1)$ quota. Accumulated thickness of adsorbed gas layer in contacting zone itself comes to $\Theta_\Sigma = 1$ under pressure action. The system comes to an adhesive friction state in this contacting zone. The friction system comes into adhesive friction state at $\Theta_\Sigma = 1$, where the friction coefficient reaches its maximum, as is shown in Figure 3.10.

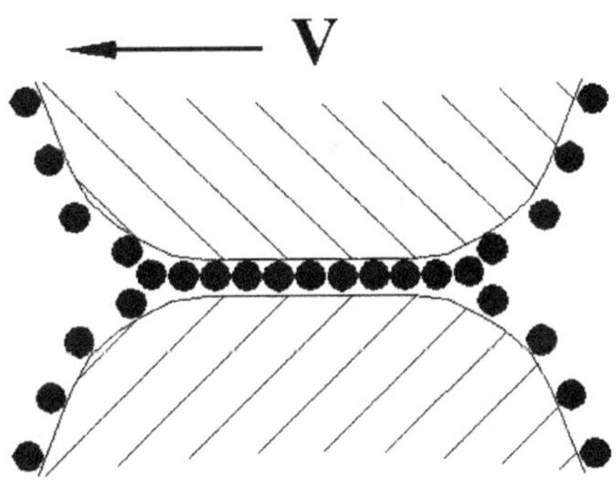

Fig. 3.11 Scheme of adhesive friction process.

3.4.4 Adhesive Friction

Adhesive friction exists in the residual pressure range, which ensures the thickness at adsorbed layer between the moving solid surfaces equal to the diameter of one adsorbed molecule $\Delta = d_0$, as it is shown in Figure 3.11.

It corresponds to the accumulated coverage range

$$1 \geq \Theta_\Sigma \geq \frac{N + F_{\mathrm{AN}}}{\sigma_{B2} \cdot A \cdot P^4(X)}, \tag{3.26}$$

where σ_{B2} is the breaking point of the weakest solid surface material in the considered friction couple.

The right part of the inequality (3.26) shows the coverage meaning, which corresponds, to the process of the adsorbed molecules implantation into the solid of the friction surface (compulsory solution) under the normal load N action.

In adhesive friction range we use energy theory also as in viscous-adhesive friction region for the friction force and for the friction coefficient calculation. The difference is that instead of the evaporation heat E_L plus melting heat E_M we use the adsorption energy E_A, because in this case the sliding plane is situated between the adsorbed gas and the solid body lattice.

The adhesive friction force may be calculated

$$F_A = \frac{W_X}{X} = \frac{E'_A \cdot A \cdot \Theta_\Sigma \cdot P^4(X)}{L_X}, \tag{3.27}$$

where E^1_A is the least specific adhesive energy "sorbate – solid" for the materials of friction couple and the adsorbed gas, $\mathrm{J \cdot m^{-2}}$.

The friction coefficient may be calculated

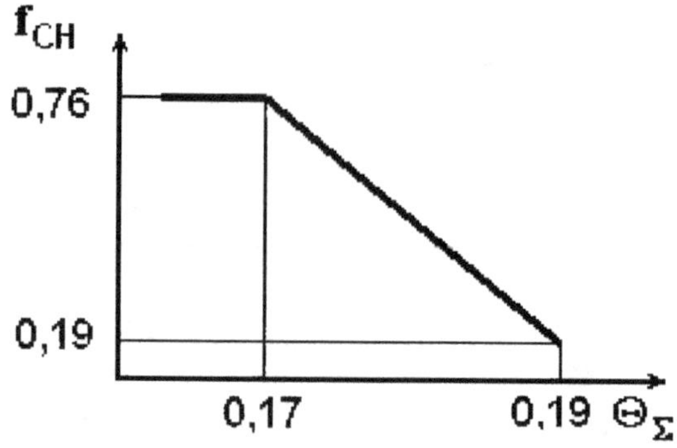

Fig. 3.12 Coefficient of adhesive friction F_A as a function of coverage coefficient for the measured sample.

$$f_A = \frac{E'_A \cdot A \cdot \Theta_\Sigma \cdot P^4(X)}{(N + F_{AN}) \cdot L_X}, \tag{3.28}$$

where F_{AN} is normal component of adhesive friction force [6], N.

After comparison of Equations (3.27) and (3.28) with Equations (3.23) and (3.24) we can see that the correlation between forces F_{AV} and F_A and friction coefficients f_{AV} and f_A corresponds to the correlation between $(E_L^1 + E_M^1)$ and E_A^1 for the considered friction couple. According to Equation (3.28) the friction coefficient linearly depends on energy $(E_A^1 \cdot A)$ and on Θ_Σ coefficient.

After quota Θ_Σ decreases to critical value

$$\Theta_{\Sigma CH} = \frac{N + F_{AN}}{\sigma_{B2} \cdot A \cdot P^4(X)} \tag{3.29}$$

the increased normal pressure acts on the squeezed adsorbed molecules and destroys the lattice of the contacting bodies. The adsorbed gases dissolve into lattices on the Θ_Σ quota of the surfaces [29] compulsory.

At $\Theta_\Sigma < \Theta_{\Sigma CH}$ the contacting surfaces come to the "real" contacting with heat energy generation E_{CH}, caused by cohesive forces.

3.4.5 Cohesion Friction

When the accumulated coverage becomes small enough

$$\Theta_\Sigma < \frac{N + F_{AN}}{\sigma_{B2} \cdot A \cdot P^4(X)} = \Theta_{\Sigma CH} \tag{3.30}$$

the adsorbed gas molecules cannot keep the normal load they were mechanically implanted into the lattice of solid contacting body with the smallest breaking point

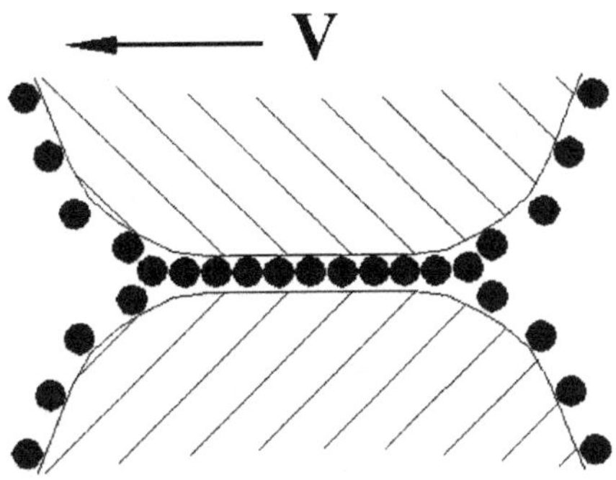

Fig. 3.13 Scheme of cohesive friction process.

σ_B that was shown by the experiments [30, 31]. Adhesive nature of friction transforms into cohesive nature of friction. Contacting surfaces of the friction couple come into "true" contact, Figure 3.13.

The vacuum which corresponds to the coverage according to Equation (3.30) may be titled "extremely high vacuum" to distinguish it from "ultrahigh vacuum" which corresponds to Equation (3.25). The described results form the basis for the term "extremely high vacuum" introduction into classical theory of vacuum [32].

The contacting molecules of the surfaces form "cohesive links" ("ion links" in metal).

Cohesive component of friction force, calculated according to the energy theory of friction may be expressed:

$$F_{\mathrm{CH}} = \frac{W_X}{X} = \frac{E_{\mathrm{CH}} \cdot A \cdot P^4(X)}{L_X}, \tag{3.31}$$

where E^1_{CH} is specific adhesive energy between the molecules of contacting solid surfaces, J·m^{-2}.

The "true" cohesive friction coefficient may be calculated if we take into account normal cohesive linking forces F_{CHN} between contacting surfaces:

$$f_{\mathrm{CH}} = \frac{F_{\mathrm{CH}} \cdot A \cdot P^4(X)}{(N + F_{\mathrm{CHN}}) \cdot L_X} \approx \frac{\tau_B}{\sigma_B}. \tag{3.32}$$

The cohesive friction coefficient does not depend on accumulated coverage Θ_Σ and on residual pressure in vacuum chamber, as is shown in Figure 3.14 in the coverage range, corresponding to inequality $\Theta_\Sigma < 0.17$.

As it was shown above the behavior of friction coefficient in different ranges of residual pressure (and in different ranges of coverage) depends on accumulated coverage variation. A graph for the behaviour of the friction coefficient calculated theoretically for the SiO_2 surfaces is shown in Figure 3.15.

In the graphs we can see four maximum values of the friction coefficient:

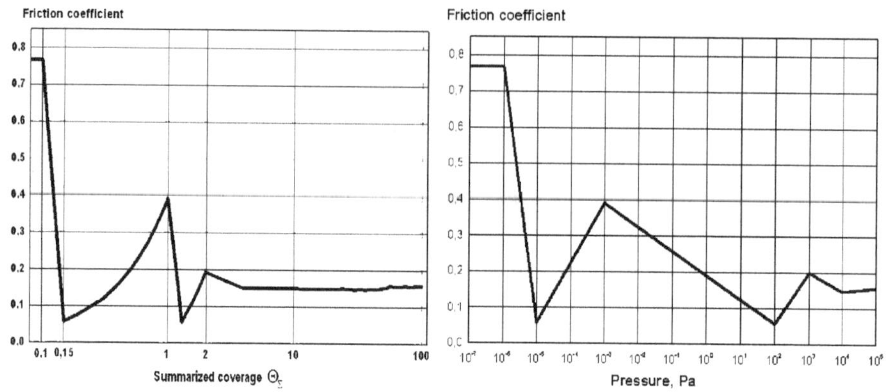

a) b)

Fig. 3.14 Coefficient of cohesive friction F_{CH} as a function of Θ_Σ, for the measured sample (see Figure 3.2).

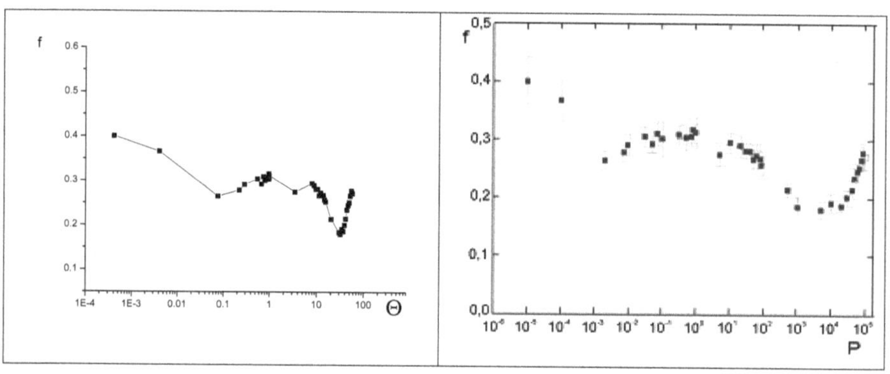

a) b)

Fig. 3.15 Variation of theoretically calculated friction coefficient (for the silicone monocrystal profile, Figure 3.2) versus: (a) summarized coverage; (b) residual pressure.

1. At $\Theta_\Sigma = 20 \ldots 150$ (at $\Theta_\Sigma = \Theta_{\max}$) because of the largest value of contacting square, which is close to geometry (nominal) contacting area.
2. At $\Theta_\Sigma = 2$ because of the maximum value of viscous friction gradient in the "quasi-liquid" part of the accumulated coverage.
3. $\Theta_\Sigma = 1$ because of the largest value of the smooth contacting area covered with adsorbed gas.
4. At $\Theta_\Sigma < (N + F_{\text{CHN}})/(\sigma_B \cdot A \cdot P^4(X))$ because of the largest (cohesive) attractive energy (energy of friction) in contacting spots.

The results of the experiments [33] which were done with the sample surface geometry (used in the above-mentioned calculations) are consistent with the model.

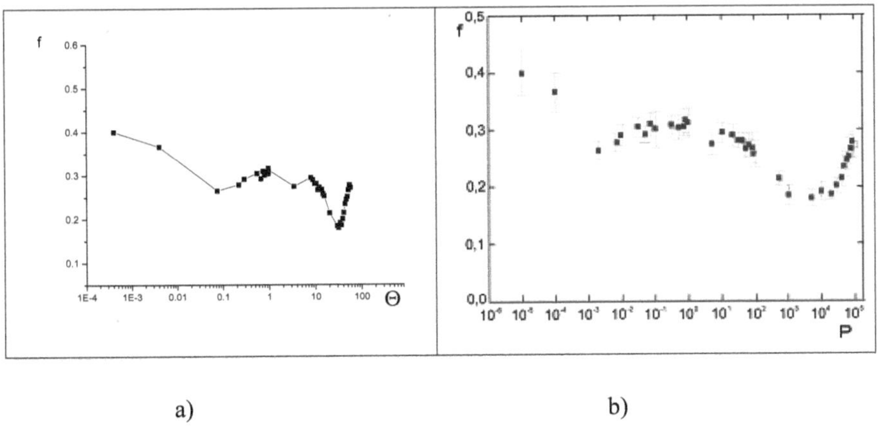

Fig. 3.16 Variation of experimentally found Si-Si friction coefficient (for the same profile) versus: (a) summarized coverage; (b) residual pressure.

3.5 The Possibility to Use the Described Method for the Calculation of the Friction Coefficient of Real Surfaces

The real "rough" surfaces are different from the "smooth" surface of a polished monocrystal Si (or SiO_2), shown in Figure 3.2, since they have no contacting spots, which may be considered as ideally smooth. Besides, they have the larger height of the micro summits.

It is obvious to suppose that in low vacuum, i.e. at coverage $\Theta \approx 150 - 2$, when the micro summits are immersed in a quasi-liquid adsorbed gas, the friction behaviour of rough surfaces will be the same as for smooth surfaces and the friction coefficient will be at its minimum at coverage $\Theta_\Sigma = 10 - 30$ (that corresponds to the partial water pressure 0.1–1000 Pa or residual accumulated pressure P = 10–10000 Pa). This proposition is well illustrated by the previously published results, where the corresponding coverage can be calculated.

So, in the results presented in Figures 3.17 and 3.18 [34], we can see the minimum values of the friction coefficients reached in the range of the pressures 10^2–10^4 Pa (depending on speed and load).

Using the parameters presented in [33], the surface coverage was calculated. Figure 3.17 shows that in correspondence with Equations (3.9), the contacting load increasing from 0.5 N/m^2 to 2.0 N/m^2 leads to a shift of the border ($f_{AV} - f_{CV}$) between adhesive-viscous and capillary-viscous regimes from the pressure $5 \cdot 10^{-3}$ Pa to the pressure $1.5 \cdot 10^{-1}$ Pa.

Figure 3.18 shows that in correspondence with Equation (3.9), when the sliding speed increases from 0.18 m·s^{-1} to 0.337 m·s^{-1} leads to a shift of the border ($f_{AV} - f_{CV}$) between adhesive-viscous and capillary-viscous regimes to the right from the pressure $2 \cdot 10^{-3}$ Pa to pressure 10^{-1} Pa. The border shift is illustrated by the slanted line, which corresponds to the border between f_{AV} and f_{CV}.

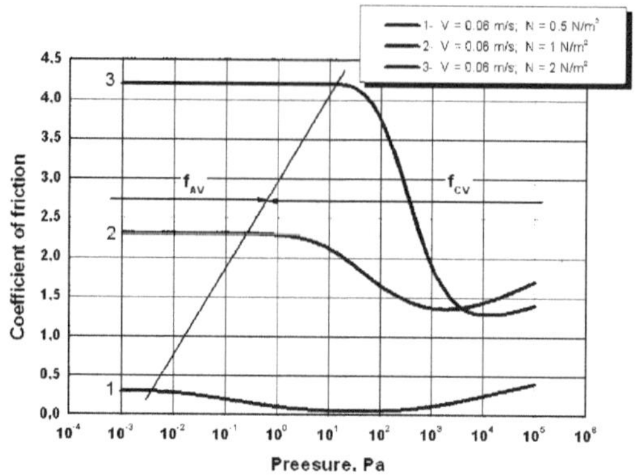

Fig. 3.17 Friction coefficient variation for "carbon steel–carbon steel" friction couple as a function of residual pressure and load at constant friction speed [33].

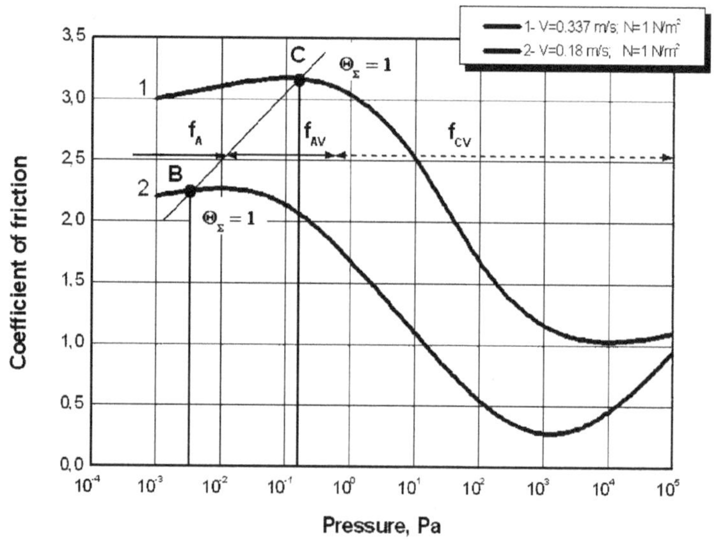

Fig. 3.18 Friction coefficient versus residual pressure for "carbon steel–carbon steel" friction couple [33] for different friction speed.

The results of the investigation of ball bearings in vacuum [35] are presented in Figure 3.19. A ball bearing is the most typical friction mechanism. These results illustrate a (sliding and rolling) friction of "rough" surfaces and also show that the presented theoretical model fits well. We can see some extremes in the presented graph, which correspond to the friction nature variation, as it was shown above. The left (basis) graph shows the experimentally obtained results of the torque variation as a function of the residual pressure. The right graph of Figure 3.19 shows the same results presented in the coordinates "torque – coverage". We can see from the

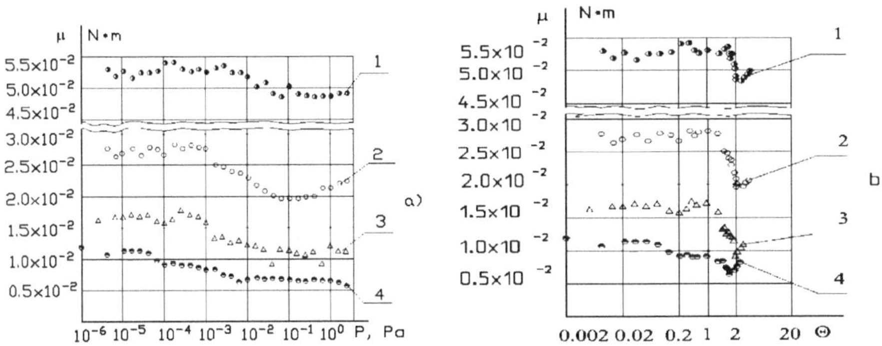

Fig. 3.19 The results of the investigation of ball bearing in vacuum [34]: the left (basis) graph shows the experimentally obtained results of the torque moment variation as a function of the residual pressure, the right graph shows the same results presented in the coordinates "torque moment–coverage".

graph that the border ($f_{AV} - f_{CV}$) between adhesive-viscous and capillary-viscous regimes in this case corresponds to the coverage $\Theta = 1.4$, which was calculated for maximum "differential" sliding speed of the balls with the trace.

The most interesting results we found after coordinate "torque" transformation in Figure 3.19 into coordinate "friction coefficient", as shown in Figure 3.20a. All experimental curves for different ball bearings tested at different loads and speeds of rotation, Figure 3.19, are joined into one graph, which is similar to the graph of silicon behaviour, (as the theory predicts), Figure 3.15a and experimentally measured, Figures 3.16a and 3.20b.

The presentation of the other recalculated results of different authors [36, 37] shows that all results, presented in the coordinates "friction coefficient – coverage" are similar to the results presented in Figure 3.15a, 3.16a and situate in the same area of the graphs.

So, Figure 3.21a shows the original (initial, before recalculation) results [38] of SiO_2–SiO_2 friction investigation and Figure 3.21b shows the same results but presented in recalculated coordinates (presented in coordinates "friction coefficient – coverage"). We can see that in this coordinate system the analyzed results are very close to the ones presented in Figures 3.15a and 3.16a.

With the idea about the unity of differently presented friction results in mind, let us consider Figure 3.22a, which shows the original (initial for recalculation) results of "Si-sapphire" friction at different relative humidity RH [35]. The first idea that comes to mind, is that the results in Figure 3.22 are unrelated to the results of Figure 3.16, but after the corresponding coverage calculation, Figure 3.22b, we can see that these are the same results, but presented in recalculated coordinates (presented in coordinates "friction coefficient – coverage"). As in the previous case, we can see that in this coordinate system the analyzed results are very close to other ones, as presented in Figures 3.15a and 3.16a.

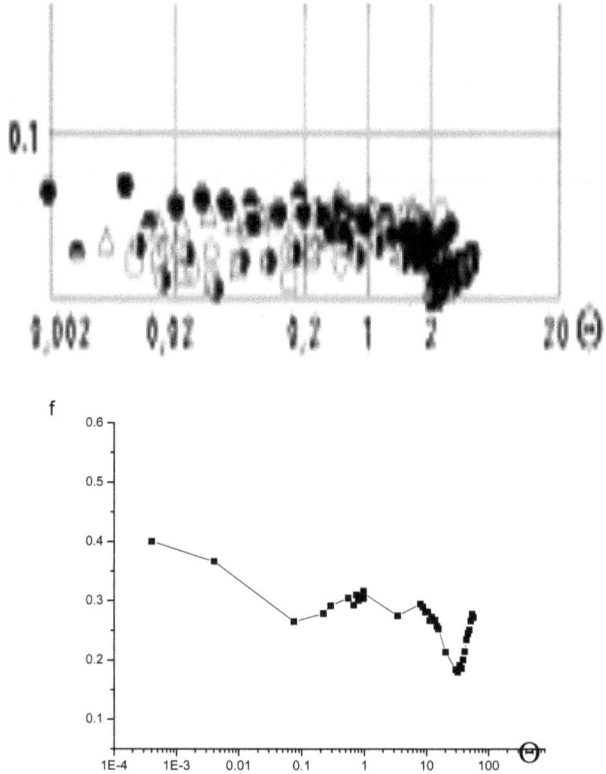

Fig. 3.20 The resulting view of the presentation of the experimentally obtained results in coordinates "friction coefficient – coverage": (a) ball bearing [34]; (b) silicone monocrystal–silicone monocrystal [20].

For consolidating the above-mentioned idea, let us consider Figure 3.23, where most of the above-mentioned results are presented in one graph. We can see that all results, presented in the coordinates "friction coefficient – coverage" situate in the same area of the graph and the behaviour of friction coefficients determined under different conditions in the different areas of the graph is similar. The difference of the curves is only found in the values of the friction coefficient, because of the difference of the tested materials (SiO_2–SiO_2 [36], Si-Si hydrophilic and hydrophobic-sapphire [35], Si-Si hydrophilic [20, 38]).

Thus, we can see that the dry friction process is a complex process which depends on the coverage of the surface as well as on the energy spent (or generated) in the contacting zone. This dry friction process consists of four different components, which dominate in a certain region of coverage:

1. Capillary-viscous friction acts in low and middle vacuum (at total coverage $\Theta_\Sigma \geq 2$);

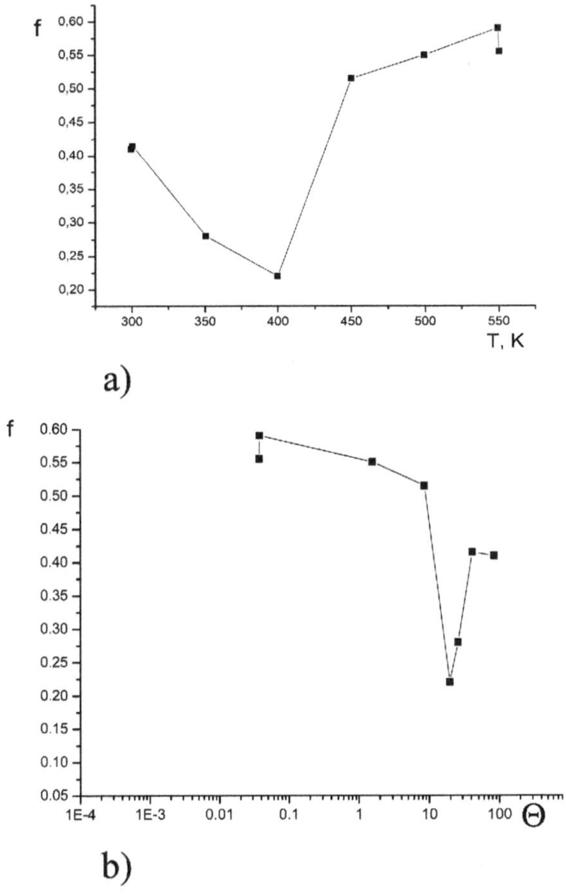

Fig. 3.21 Experimental results of SiO_2-SiO_2 friction as function of the sample temperature [36]. (b) The same results presented in coordinates "friction coefficient – surface coverage" (in recalculated coordinates).

2. Adhesive-viscous friction in high vacuum at $2 > \Theta_\Sigma \geq 1$.
3. Adhesive friction in ultrahigh vacuum (at $1 > \Theta_\Sigma > \Theta_{\Sigma CH}$).
4. Cohesive friction acts in extremely high vacuum ($\Theta_\Sigma < \Theta_{\Sigma CH}$).
5. New experimental results as well as previous results of other authors fit well into the presented theory.

3.6 Exchange of Gases at Friction in Vacuum

It was shown that among all residual gases adsorbed on the surfaces water vapour is the most important to determine the behaviour of the friction coefficient. The

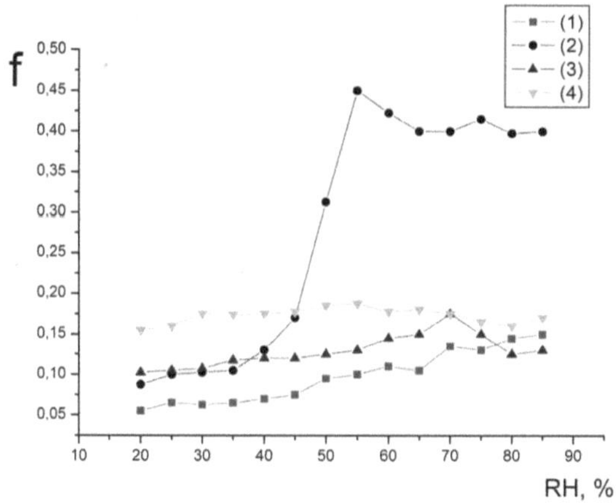

a)

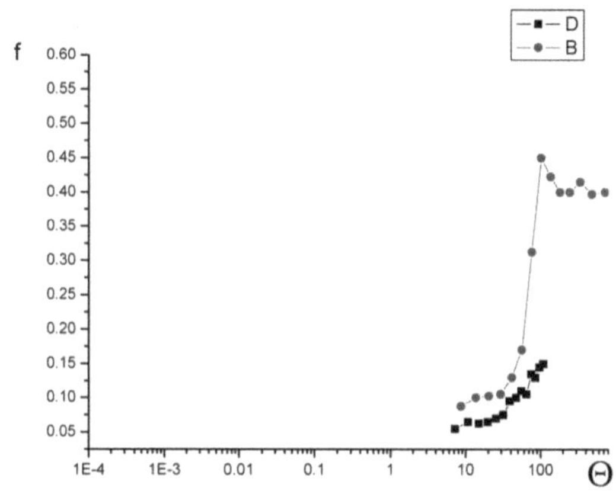

b)

Fig. 3.22 (a) Experimental results of Si-sapphire friction at different relative humidity RH [35]: 1. hydrophilic Si-sapphire, $t = 25°C$, 2. hydrophilic Si-sapphire, $t = 80°C$, 3. hydrophobic Si-sapphire, $t = 25°C$, 4. hydrophobic Si-sapphire, $t = 80°C$. (b) The data for hydrophilic Si presented in coordinates "friction coefficient – coverage" (in recalculated coordinates).

purpose of the study presented below is to show the source and the mechanism of H_2 and D_2 dissolution in the materials of friction pairs and to describe the kinetics of its accumulation in surface layers.

Now we try to show that the main source of the increased hydrogen and deuterium concentration in the contacting materials is water, which is adsorbed on the surface. We consider that dissolution of adsorbed gases takes place both on free sur-

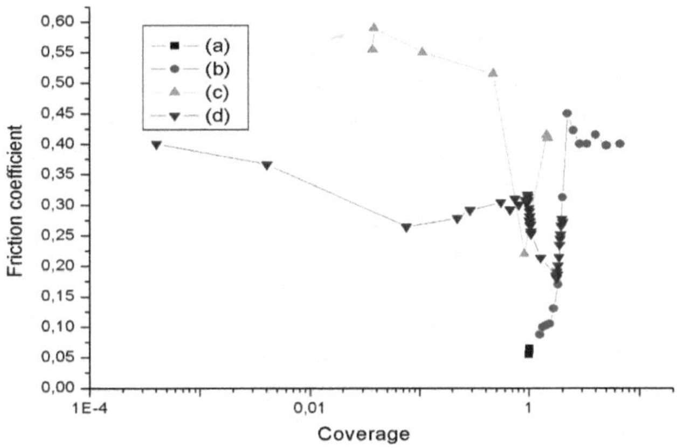

Fig. 3.23 United graph presenting the different results in the coordinates "friction coefficient – coverage": (a) hydrophilic Si-sapphire, $t = 25°C$, tested at different RH [35]; (b) hydrophilic Si-sapphire, $t = 80°C$, tested at different RH [35]; (c) SiO_2-SiO_2 tested at different temperatures [36]; (d) hydrophilic Si-hydrophilic Si, tested at different pressures [37].

face at keeping a sample in the natural atmosphere and in rather higher degree on the contact surfaces under friction.

The calculation has shown that the residual pressure of the main dissolving gases, i.e. hydrogen, is too small to initiate the dissolution process. It does not correspond to the equilibrium state of the dissolved gases. The equilibrium amount of a gas dissolved in a solid body can be expressed [39]:

$$S = S_0\, p^{1/j} \exp\left(\pm \frac{E_s}{j R_0 T}\right) \; [\mathrm{m^3\, Pa\, m^{-3}}], \tag{3.33}$$

where S_0 is constant of solubility, $[\mathrm{Pa}^{1/j}]$ ($S_0 = 2 \cdot 10^3\ \mathrm{Pa}^{0.5}$ for H_2 in AISI 304 steel, $S_0 = 1.85 \cdot 10^3\ \mathrm{Pa}^{0.5}$ for D_2 in AISI 304 steel; p is the residual pressure of the gas being dissolved in a solid, [Pa]; E_s is the activation energy of solubility, $[\mathrm{J{\cdot}mole^{-1}}]$ ($E_s = 1.4 \cdot 10^4\ \mathrm{J{\cdot}mole^{-1}}$ for H_2 in AISI 304 steel, $E_s = 1.51 \cdot 10^4\ \mathrm{J{\cdot}mole^{-1}}$ for D_2 in AISI 304 steel; j is the number of atoms in a molecule; T is the temperature [K]; R_0 is the gas constant $[\mathrm{J{\cdot}mole^{-1}{\cdot}K^{-1}}]$, ($R = 8.31\ \mathrm{J{\cdot}mole^{-1}{\cdot}K^{-1}}$).

The dissolved gas concentration may be expressed as atom concentration:

$$C = S(kT)^{-1} \; [\mathrm{m^{-3}}], \tag{3.34}$$

where k is Boltzmann constant, $[\mathrm{J\,K^{-1}}]$ ($k = 1.38 \times 10^{-23}\mathrm{J\,K^{-1}}$).

The calculation results of the residual H_2 and D_2 dissolution in the stainless steel have shown that at the residual H_2 atmospheric partial pressure $P_{H_2} = 5 \cdot 10^{-2}$ Pa its equilibrium concentration in stainless steel should be $6.5 \cdot 10^{15}$ at cm^{-3}. Correspondingly, at the residual D_2 atmospheric partial pressure $P_{D_2} = 7 \cdot 10^{-6}$ Pa its equilibrium concentration in stainless steel should be $8 \cdot 10^{13}$ at cm^{-3}.

In [40, 41] it was shown that water adsorbed on the surfaces dissociates into H^+ (or D^+) and OH^- (or OD^-) ions, which become adsorbed on the surfaces. The increased concentration of the adsorbed (hydrogen and deuterium) molecules on the surface is a main initial condition for its dissolution in the surface layers of friction pairs. The SIMS analysis has shown [29] that near surface there is an increased concentration of these gases in reference samples before friction. This proves that the dissolution processes of these gases on the surface does take place, but less intensively than during the friction process.

The test results $6 \cdot 10^{18}$ at cm^{-3} for hydrogen and $(0.8–4) \times 10^{17}$ at cm^{-3} for deuterium, i.e. the concentration of the naturally dissolved gases H_2 and D_2, are 1000 times higher than follows from (3.33) and (3.34). We consider that friction stimulates the dissolution process of adsorbed gases, because heat, photons, ions and electrons are being generated [42].

Let us discuss the adsorbed gases behaviour during the friction process. Both in vacuum and in atmosphere five main stages can be considered for gas-solid reactions on the contact surfaces at friction contacts.

Gas-surface interaction during friction depends on temperature distribution and distribution of gas concentration on the surface as well in subsurface volume. Initial surface temperature is considered to be equal to initial gas temperature. Initial gas concentration in material of friction couple C_0 corresponds to the concentration of gas in material after outgassing baking.

Initial gas concentration on contact surfaces depends on pressure p, temperature T_s, adsorption energy E_a and during the friction process it is determined by surface coverage θ, see Equations (3.9) and (3.9a).

Let us estimate the gas concentration in the different parts of considered system:

- *Concentration of gases above the surface* is equal to the concentration of residual gas. For $p = 10^{-3}$ Pa the concentration above the surface is $C_p = 2.7 \cdot 10^{17}$ at m^{-3} ($C_p = 10^{-3}$ Pa·m^3/m^3).
- *Concentration of gases on the surface* depends on the coverage. In the case $\Theta > 1$ the surface-gas concentration is equal to the molecules concentration in a lattice of a solid. In this case the coverage is considered as a part of a solid body. For the residual hydrogen at the pressure $p = 10^{-3}$ Pa ($\Theta > 1$), the calculation gives initial volume concentration of adsorbed gas on surface $C_s = 5.53 \cdot 10^{28}$ m^{-3} (or in PV-units $C_s = 2 \cdot 10^8$ Pa·m^3/m^3).
- *Concentration of gases in the lattice* varies during the friction process as it is shown below. The initial concentration of the dissolved hydrogen depends on the material treatment (for example: for the heated stainless steel the initial hydrogen concentration is $C_0 = 1.1 \cdot 10^{25}$ m^{-3}, or the same in PV units $C_0 = 4 \cdot 10^4$ Pa·m^3/m^3 [43]).

Let us consider the scheme of two micro roughnesses, coupled during the friction process (as shown in Figure 3.24). Depth profiles of temperature and gas concentration distributions in subsurface layers are shown in Figure 3.25 (left and right respectively). For the considered example the initial temperature is $T_{si} \approx 293$K, initial volume gas concentration in material $C_0 \approx 10^4$ Pa·m^3/m^3, volume gas con-

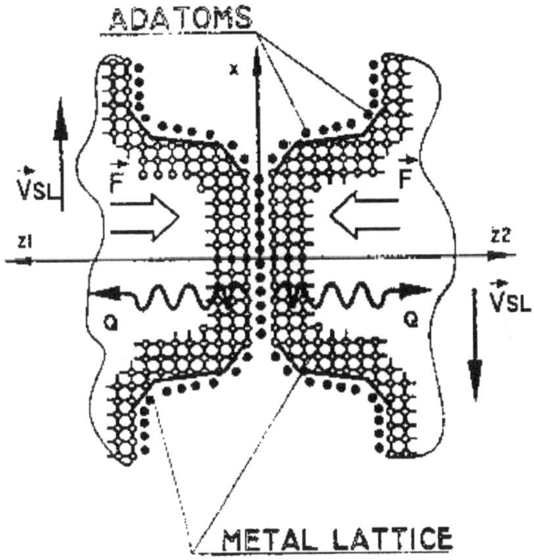

Fig. 3.24 Scheme of coupling of two micro roughness at friction: F – load; Q – heat flow; V_{sl} – sliding speed.

centration on surface Cs $\approx 10^8$ Pa·m^3/m^3 and volume gas concentration (in vacuum volume) $C_p \approx 10^{-3}$ Pa·m^3/m^3 (Figure 3.25a).

At the moment, when the contact begins (Figure 3.25b) a heat, resulting from the generation of the specific evaporation energy E'_L, J·m^{-2}, or specific melting energy E'_M, J·m^{-2} (see Equation 3.21) is accumulated and the temperature of the adsorbed atoms, squeezed between surfaces, rises instantly up to the level 2 (Figure 3.25b, left).

This heat is transferred on to metal lattice atoms and to squeezed fixed gas atoms. Part of this energy is spent for the excitation of the molecules and their dissociation, for the different kinds of radiation [45] and that is why this part of the generated energy may be considered as activation energy of diffusion E_{ad}.

Initial depth distribution of volume gas concentration for the beginning of the friction process is shown in Figure 3.25b (right). It corresponds to the initial conditions pointed in Figure 3.25a (right).

Temperature in subsurface layers is varying during coupling. Distribution of temperature is determined both by thermal conductivity of metal and by heat generation on the contacting surfaces during sliding friction (levels 3–5), Figure 3.25c (left).

Concentration of gas, diffusing from the contact area where gas concentration is the highest, may be evaluated by non-stationary Fick's equation.

The temperature in the middle of a mating solid body, which is formed by the contacting surfaces at every short moment limited by coupling process increases continuously, and the concentration of quasi dissolved squeezed gas atoms decreases

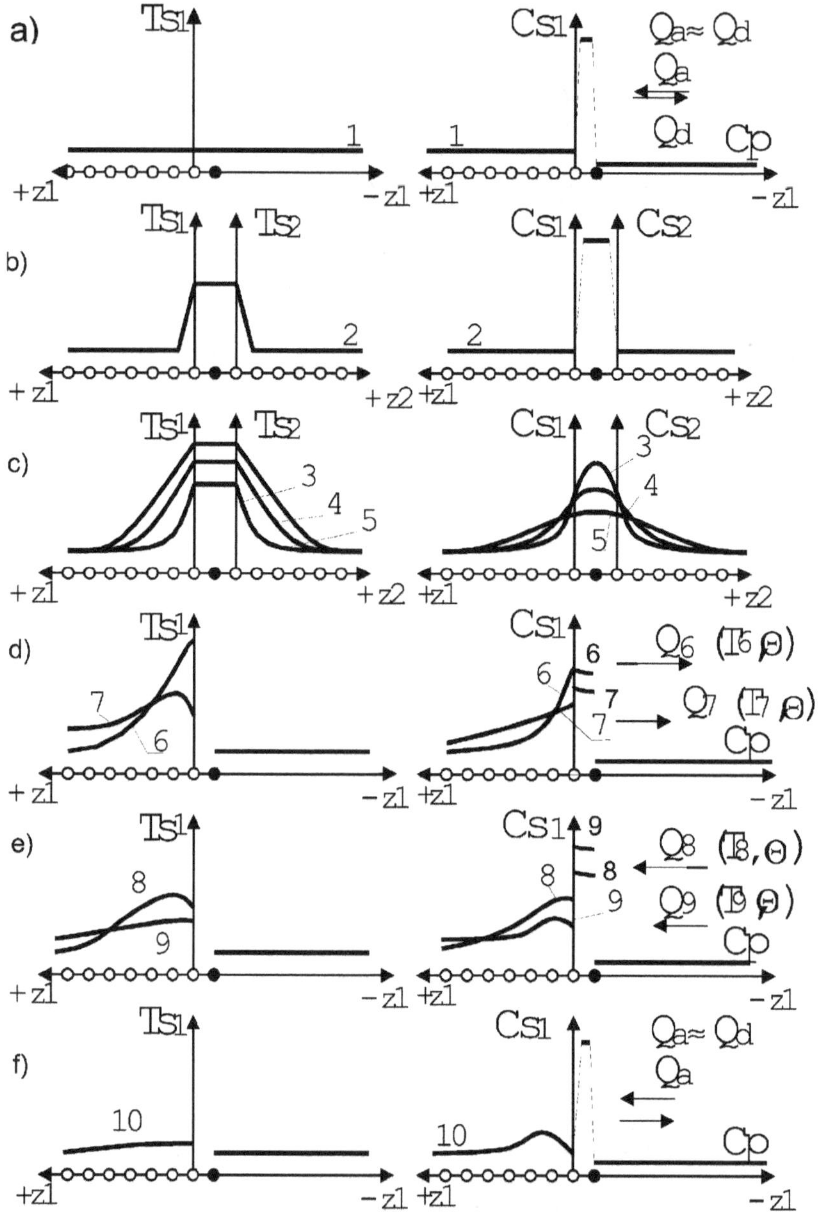

Fig. 3.25 Curves of temperature distribution in depth (left) and of gas concentration distribution in depth (right). z_1, z_2 – distance from surface for the first and the second coupling bodies respectively; T_{s1}, T_{s2} – depth distribution of temperature in the first and second bodies, respectively; C_0, C_s, C_p – gas concentration in material, on surface and in vacuum volume respectively; Q – total specific gas flows; stages of friction process: (a) initial state; (b) beginning of contact; (c) process of friction; (d) outgassing; (e) gas gettering on the surface; (f) resulting equilibrium.

during coupling because the number of the adsorbed atoms, squeezed between surfaces is limited and decreases due to diffusion into metal lattice (levels 3–5, right).

When coupling is finished (Figure 3.25d), surfaces are facing into vacuum again. Gas temperature in vacuum chamber is constant ($T = 293$K in our example) but temperature of coupling surfaces has become high and in subsurface layers it is distributed in an exponential way with maximum value on surface (the estimation in our example according to [1] gives $T \approx 400$ K).

After the contacting process is finished, a concentration of dissolved gas atoms in the subsurface layers is higher than the equilibria are for this high temperature. So, it starts to decrease, due to active gas release Q_6. Simultaneously, at high temperature, the process of diffusion of dissolved molecules from the surfaces, i.e. from the place of the highest concentration, into the solid takes place.

High temperature of the open surface decreases rapidly due to heat radiation, due to metal heat conductivity and due to expenditure of the energy for the gas release (of the molecules fixed on the surface). Necessary to note that the biggest part of atoms and molecules on the surface did not dissolve in the contacting process. Temperature in subsurface layers adapts due to heat conductivity of metal. (At some distance from the surface it continues to rise, levels 6, 7.)

Decreasing of temperature and gas concentration at a high rate corresponds to decreasing of gas release flow according to Equations (3.7) and after some time gas desorption vanishes. System "surface-residual gas" comes to the equilibrium.

Further cooling of the surfaces after contacting causes disturbance of the equilibrium state described by (3.8) and (3.9) that stimulates filling of adsorption vacancies, formed on the surface at stage "d". It means that the adsorption begins to dominate over desorption (Figure 3.25e, right, levels 8, 9) and the couple of surfaces begins "to pump" gas, until the system gas-surface comes to equilibrium state (Figure 3.25f).

It is necessary to remember that all discussed processes take place in microvolumes (depth about 100 nm) at a very short period of time (for example $t \leq 50$ ms for ball bearings at a speed of rotation: 200 rpm).

Let us consider the experimental confirmation of the presented model.

The results of the experimental research of the mechanically stimulated hydrogen solution in a natural atmosphere [44] are shown in Figure 3.26. The special sliding friction device was mounted on the system described in [29] and it was used instead of the rotary-motion feedthrough. In this case the flat sample was used. Maximum contacting strain was $\sigma = 1500$ MPa, contacting frequency $n = 60\ \text{s}^{-1}$, cumulative number of contacting $N = 6000$.

Analysis of the results after mechanical action shows the increase in hydrogen concentration at the depth of 6 microns in comparison with the reference sample. Maximum concentration of the dissolved hydrogen $C_H^{\max} = 2 \cdot 10^{21}$ at·cm^{-3} is situated at the depth of 0.2 microns and this concentration is 300 times higher than the hydrogen concentration in the reference sample ($C_H = 6 \cdot 10^{18}$ at·cm^{-3}).

Hydrogen concentration distribution in the reference sample shows great decrease of hydrogen concentration (from $4 \cdot 10^{19}$ to $6 \cdot 10^{18}$ at·cm^{-3}) at the depth of 0.15 microns from the surface and shows small decreasing of concentration (from

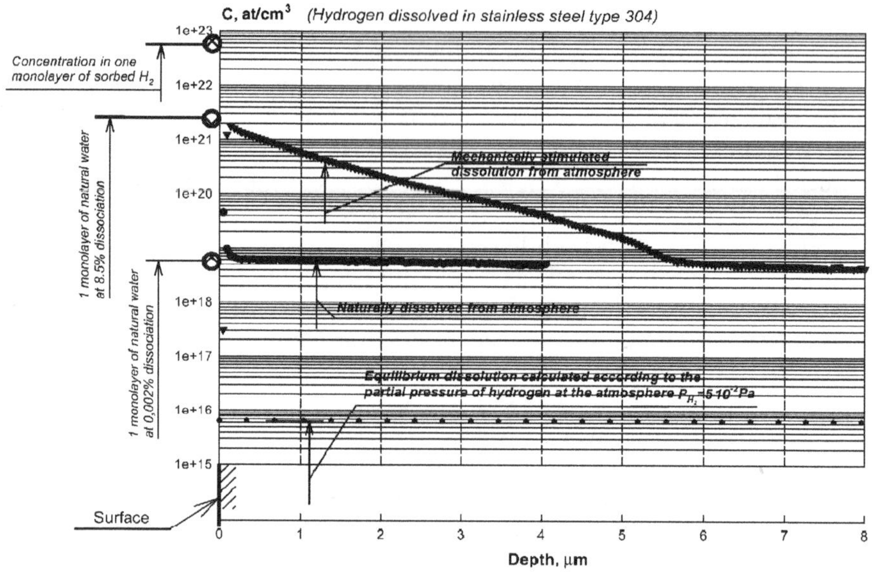

Fig. 3.26 Dissolved hydrogen concentration distribution in the flat sample after friction in the natural atmosphere (upper curve $\sigma = 1500$ MPa, $N = 6000$ contacts, $P_{H2} = 5 \cdot 10^{-2}$Pa) and in the reference sample (lower curve, $N = 0$, $P_{H2} = 5 \cdot 10^{-2}$ Pa).

$6 \cdot 10^{18}$ to $5 \cdot 10^{18}$ at·cm^{-3}) at the depth of 0.15 microns to 4.0 microns. This behavior of the distribution leads us to think that for both samples there is a constant process of hydrogen diffusion from the surface into the solid body of the sample, but this process is intensified by friction.

In the experiment with deuterium [29, 30] the ball of a ball bearing of the rotary motion feedthrough was used as a sample. Ball bearing was rotated in the natural atmosphere with rotation speed $n = 200$ r.p.m. and it was loaded with an axial load $A = 240$N and radial load $B = 50$N, maximum contacting strain was $\sigma = 1740$ MPa. The accumulated number of contacting cycles was about $N = 7.5 \cdot 10^4$.

The results of this experiment show the deuterium concentration distribution, Figure 3.27, in the sample after friction (upper curve) and in the reference sample (lower curve). Maximum deuterium concentration after friction ($C_D^{\max} = 4.5 \cdot 10^{19}$ at·cm^{-3} at the depth $\delta = 0.05$ micron) is 50 times higher than the deuterium concentration in reference sample ($C_D = 9 \cdot 10^{17}$ at·cm^{-3} at the depth $\delta = 0.05$ microns). We can see the decreasing of the deuterium concentration to the depth of 0.9 micron that can be explained by the constant processes D diffusion from atmosphere.

Deuterium concentration distribution in the reference sample shows large decreasing of the concentration (from $3 \cdot 10^{18}$ to $6 \cdot 10^{17}$ at·cm^{-3}) at the depth $\delta = 0.05$ micron and it shows also almost constant distribution and a large dispersion of the concentration of C_D, $C_D = (6 \cdot 10^{18}$ at·cm^{-3}-$7 \cdot 10^{17}$ at·sm^{-3}) at the depth from $\delta = 0.05$ to 1.0 micron. We can see also that the distribution of the

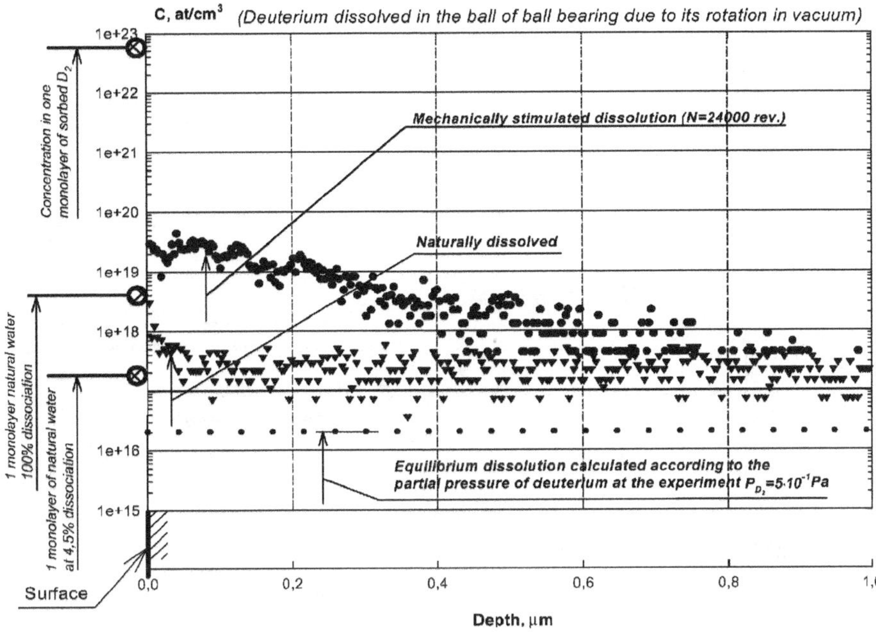

Fig. 3.27 Dissolved deuterium concentration distribution in the sample (ball made of steel type 304) after friction in the natural atmosphere (upper curve $N = 75000$ contacts, $\sigma = 1740$ MPa, $P_{D2} = 7 \cdot 10^{-6}$ Pa) and in the reference sample (lower curve $N = 0$, $P_{D2} = 7 \cdot 10^{-6}$ Pa).

deuterium concentration is not as stable as the hydrogen one, but the behaviour of the concentration distribution leads us to think this is caused by the constant process of deuterium diffusion from the surface into the solid lattice of the sample.

The purpose of the third experiment was to find the source of hydrogen and deuterium penetrating into the bulk of the samples from the surfaces.

The stainless steel (type 304) sample was placed into a deuterium atmosphere $P_{D2} = 5 \cdot 10^4$ Pa in a sealed volume. Sample was sealed with a distillate water gate and it was kept in a deuterium atmosphere for 76 days. Figure 3.28 shows the results of this experiment. We can see that in the sample exposed for 76 days in a deuterium atmosphere ($P_{D2} = 5 \cdot 10^4$ Pa) the maximum deuterium concentration ($C_D^{\max} = 7 \cdot 10^{17}$ at·cm^{-3}) becomes only 100 times higher than the deuterium concentration in the reference sample ($C_D = 1 \cdot 10^{16}$ at·cm^{-3}) kept in a natural atmosphere ($P_D = 7 \cdot 10^{-6}$ Pa). According to Silvert's law [38] and Equation (3.33) the deuterium pressure increase 10^{10} (from $P'_{D2} = 7 \cdot 10^{-6}$ Pa to $P''_{D2} = 5 \cdot 10^4$ Pa) must be accompanied by an increase of concentration in $(P''_{D2} : P'_{D2})^{0.5}$, i.e. approximately 10^5 times.

It shows that there is some kind of barrier on the surface, which limits the adsorption and dissociation processes of deuterium on the surface and it decreases them by approximately 10^3.

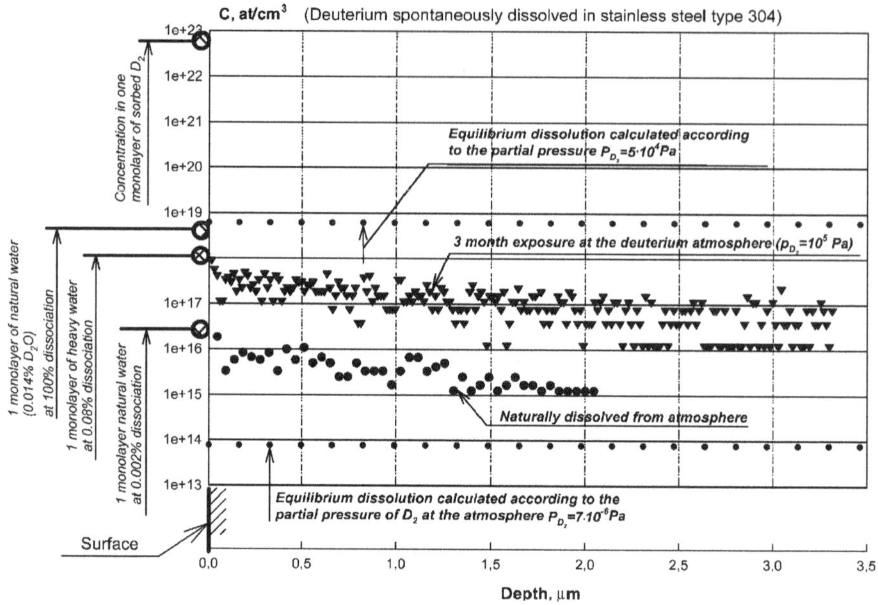

Fig. 3.28 Dissolved deuterium concentration distributions in the sample made of stainless steel type 304 after its long term exposure to deuterium atmosphere (upper curve) and in reference sample kept in natural atmosphere (lower curve).

On the other hand, this experiment shows the presence on the surface the source of deuterium, and shows also that the deuterium concentration in this source is much higher than an equilibrium concentration of deuterium in natural atmosphere. Calculation of the equilibrium deuterium concentration for its natural partial pressure ($P_{D2} = 7 \cdot 10^{-6}$ Pa) gives dissolved concentration $C_D = 8 \cdot 10^{13}$ at·cm^{-3}, which is 100 times less than the experimentally found maximum concentration $C_D \approx 1 \cdot 10^{16}$ at·cm^{-3}. It shows that there is some source of H and D atoms on the surface, which increases their dissolution process approximately 10^2 times.

Both of these results show that the most probable source of dissolved H and D is water on the surface. According to [39], a large part of fixed water molecules dissociates into OH^- and H^+ ions or OD^- and D^+ ions.

Experimental results presented in Figure 3.28 show that the concentration of H^+ ions as well as D^+ ions is much higher (~100 times) than the calculated concentration according to [45,46] of fixed molecules of corresponding residual gases (H_2 and D_2), see Table 3.1. On the other hand, the sample surface is completely covered with the fixed H_2O, D_2O, HDO molecules and OH^+, OD^+ ions. While residual pressure of H_2 and D_2 is increasing (even till $5 \cdot 10^4$ Pa) this can not cause the increase of the corresponding atoms H or D coverage.

Table 3.2 Theoretically calculated and experimentally measured concentrations of the dissolved H_2 and D_2 in stainless steel (type 304) without friction.

Gas	Temperature of the sample, K	Partial pressure, Pa	Theoretical concentration, at/cm^3	Experimental concentration, at/cm^3
Deuterium	300	$7 \cdot 10^{-6}$	$8 \cdot 10^{13}$	$1 \cdot 10^{16}$
Deuterium	293	$5 \cdot 10^{4}$	$6 \cdot 10^{18}$	$7 \cdot 10^{17}$
Deuterium	300	$5 \cdot 10^{-1}$	$2 \cdot 10^{16}$	$9 \cdot 10^{17}$
Hydrogen	293	$5 \cdot 10^{-2}$	$6 \cdot 10^{15}$	$5 \cdot 10^{18}$

We can also see that surface friction increases the process of dissociation of adsorbed water approximately 70–100 times. Calculation shows that natural dissociation of naturally adsorbed water is about 0.002%.

The main task of the experiments was to compare the surface concentration distributions of deuterium and hydrogen in the stainless steel type 304 after friction and after long term exposure in the residual atmosphere of deuterium with the deuterium and hydrogen concentration distributions in the reference samples.

Distribution of the concentration of naturally dissolved deuterium (Figure 3.6, lower curve) shows that the process of constant penetration of deuterium into solid bulk takes place, because the concentration decreased from the surface ($C_D = 1 \cdot 10^{16}$ to $3 \cdot 10^{15}$ at·cm^{-3}, $\delta = 0.01$ microns) and moved into the bulk of the sample ($C_D = 1 \cdot 10^{15}$ at·cm^{-3} at depth $\delta = 2.0$ microns).

The presented concept is a powerful instrument for explanation of a number of facts, which has not been explained before. These facts are:

1. The main source of hydrogen and deuterium, dissolved in solid bulk of stainless steel samples is fixed water, dissociated on the surface.
2. Surface friction stimulates the dissociation process of water about 100 times.
3. Subsurface H and D atom concentration corresponds to 0.002% dissociation of water.
4. Residual pressure of H_2 and D_2 variation causes the concentration of the dissolved H or D atoms variations, but Silvert's law does not work properly, because the water coverage plays the role of a barrier.
5. Dissolved H and D concentration distributions show that the process of constant penetration of these gases into a lattice with a small rate takes place without friction, but friction increases this process many times.
6. The amount of pumped gas is limited by small subsurface volume in the friction zone.

References

1. Kragelsky I.V., Dobychin M.N., Combalov V.S., *Foundation of Friction and Wear Calculation*, Machinostroenie, Moscow, 1977 526 pp. [in Russian].

2. Kragelsky I.V., Mikhin N.M., *Theory of Friction and Wear*, Nauka, Moscow, 1965, pp. 30–34 [in Russian].
3. Buckley D.H., *Friction, Wear and Lubrication in Vacuum*, NASA Lewis Reserch Center, 1971, Washington, 135 pp.
4. Kragelsky I.V., Lubarsky I.M., Gusliakov A.A. et al., *Friction and Wear in Vacuum*, Machinostroenie, Moscow, 1973, 215 pp. [in Russian].
5. Nusinov M.D., Vacuum simulation in the laboratory investigations of ball for the outer space use, in *Abstracts 3rd Internat. Conf. on Space Techn.*, Rome, Italy, 1971.
6. Kragelsky I.V., *Friction, Wear, and Lubrication*, Vol. 1, Machinostroenie, Moscow, 1978, 400 pp.
7. Kragelsky I.V., *Friction, Wear, and Lubrication*, Vol. 2, Machinostroenie, Moscow 1979, 632 pp.
8. Deulin E.A., Mednikov S.I., Papko V.M., *Design and Spatiality of Exploitation of the Vacuum Mechanisms*, Mashinostroenie, Moscow, 1986, 80 pp. [in Russian].
9. Dushman S., *Scientific Foundations of Vacuum Technique*, John Wiley & Sons, London, 1962.
10. Deulin E.A., Concept of dry friction of smooth surfaces in UHV, in *Proceedings 14th International Vacuum Congress (IVC-14)*, Birmingham, Abstract Book, 1998, p. 310.
11. Deulin E.A., *Mechanics and Physics of Precise Vacuum Mechanisms*, Vol. 1, Vladimir State University, Vladimir, 2001, 176 pp.
12. Vakilov A.N., Mamonova M.V., Prudnikov V.V., *Physic of Solids* **34**(4), 1997, 964–967 [in Russian].
13. Griaznov B.T., Zinkin A.N., Prudnikov V.V., Stasenko V.T., *Technological Methods of the Longevity Increasing Microcryogen Installation*, Nauka, Novosibirsk, 1999, 272 pp.
14. Volchkevitch A.I., *High Vacuum Adsorption Pumps*, Mashinostroenie, Moscow, 1973, 158 pp.
15. Dubinin M.M., Modern theory of volume adsorption for adsorption of gases and vapors on carbon sorbents, *J. Phys. Chem.* **XXXIX**(6), 1965, 1305–1317 [in Russian].
16. Brunauer F., P.H. Emmett P.H., E. Teller E., *J. Amer. Chem. Soc.* **60**, 1938, 309.
17. Bingelli M., Mate C.M., Influence of water vapor on nanotribology studied by friction force microscopy, *J. Vac. Sci. Technol. B* **13**(3), May/June, 1955, 1312–1315.
18. Archard J.F., The temperature of rubbing surfaces, *Wear* **2**, 1958/1959, 438–455.
19. Furey M.J., Surface temperatures in sliding contact, *ASLE Transactions* **7**, 1964, 133–146.
20. Dayson C., Surface temperatures at unlibricated sliding contacts, *ASLE Transactions* **10**, 1967, 169–174.
21. Deulin E.A., Gatsenko A.A., Loginov B.A., Friction force of smooth surfaces of SiO_2-SiO_2 as a function of residual pressure, *Surface Science* **433-435**, 1999, 288–292.
22. *SMM-2000T, Users Manual*, KPD Company Ltd., Moscow, 1997, 135 pp.
23. Miranda P.B., Shen L.X., Salmeron M., *Phys. Rev. Lett.* **81**(26), 1998, 5876–5879.
24. Vasiliev Y.N., et al., *Friction and Wear* **8**(6), 1987, 1044–1051 [in Russian].
25. Buckley D.H., Jonson R.I., The influence of crystal structure and some property of hexagonal metals on friction and adhesion, *Wear* **11**(6), 1968, 405–419.
26. Spurr R.T., Newcomb T.P., The friction and wear of various materials sliding against surfaces of different types and degree of roughness, in *Proceedings Conference on Lubrication and Wear*, London, Inst. Mech. Engrs., 1957, pp. 66–67.
27. Ivanova V.S., Balankin A.S., Bunin I.J., Oksogoev, A.A., *Synergetic and Fractals in Material Science*, Nauka, Moscow, 1983, 383 pp. [in Russian].
28. Feder J., *Fractal*, Plenum Press, New York, 1998, 283 pp.
29. Deulin E.A., Exchange of gases at friction in vacuum, in *Proceedings ECASIA-97*, Goteborg, John Wiley & Sons, 1997, pp. 1143–1146.
30. Deulin E.A., Goncharov S.A., Segovia J.L., Nevshupa R.A., Mechanically stimulated solution of adsorbed hydrogen and deuterium in steel, *Surf. Interface Anal.* **30**, 2000, 635–637.
31. Deulin E.A., Nevshupa R.A., Deuterium penetration into the bulk of a steel ball of a ball bearing due to its rotation in vacuum, *Appl. Surface Sci.* **144–145**, 1999, 283–286.
32. Redhead P.A., Hobson J.P., Kornelson E.V., *The Physical Basis of Ultrahigh Vacuum*, Chapman and Hall, London, 1968.

33. Gatsenko A.A., Repa P., Deulin E.A., Research of the silicon monocrystal dry friction at different vacuum degrees, *Friction, Wear, Lubrication (Electron Resource* **15**, 2002, 4 pp. [in Russian].
34. Nosovsky I.G., *Residual Gases Influence on the Wear of Metals*, Tekhnika, Kiev, 1968 [in Russian].
35. Deulin E.A., Papko V.M., Yurkov U.V., Influence of the vacuum ball bearing reliability on the output of atomized technological equipment, in *Abstracts Conference USSR BMSTU*, Moscow, 1979, pp. 38–44 [in Russian].
36. Sherge M., Li X., Schaefer J.A., *Tribology Lett.* **6**, 1999, 215–220.
37. Deulin E.A., Gladyshev I.V., Residual pressure and temperature influence on SiO_2-SiO_2 friction coefficient, in *Proceedings Nordtrib 2002, Keynotes and Abstracts*, 2000, p.176.
38. Gatsenko A.A., Řepa P., Deulin E.A., Friction coefficient variation of smooth surfaces in vacuum, in *Abstract Book of 9th Joint Vacuum Conference*, Graz (Austria), 2002, p. 20.
39. Silverts A., *Z. Metallkunde* **21**, 1929, 37.
40. Akagi K., Tsukada M., Theoretical study of hydrogen relay dissociation of water molecules on Si (001) surfaces, *Surface Sci.* **438**, 1999, 9–17.
41. Kuznetsov A.M., Water adsorption on metallic surface, *Soros Education J.* **6**(5), 45–51 [in Russian].
42. Nevshupa R.A., Nakayama K., Triboemission behavior of photons at dielectric/dielectric sliding: Time dependence nature at 10^{-4}–10^{4} s, *J. Appl. Phys.* **93**(11), 2003, 1.
43. Malev M.D., Gas adsorption and gas release of metals, *Vacuum* **23**(2), 1973.
44. Deulin E.A., Rodina E.A., Adsorbed water as a source of gases which are dissoluted in the materials, *Vacuum Technique and Technology* **13**(2), 2003, 77–82.
45. Crank J., *The Mathematics of Diffusion*, Oxford University Press, 1956.
46. Stark J.P., *Solid State Diffusion*, New York, 1978.

Chapter 4
Matrix Method of the Design of New Mechanisms Structure

As was previously shown, modern vacuum technology uses a large number of different vacuum mechanisms, and most of them are vacuum feedthroughs and drives.

The research of feedthroughs for the purpose of enhancing their precision, for the improvement of power parameters, longevity and reliability, as well as for the reduction of the outgassing rate is performed by designers simultaneously with the design of a new vacuum equipment, especially in the case, when the parameters of the existing feedthroughs do not satisfy the requirements.

The design of new mechanisms in the case of the existence of multi variables decision in combination with multi parameters limitation may be done using a matrix method for the generation of new mechanisms structure. In this method the designed object is considered as a part of a complex system with certain requirements to the object, and these requirements are compared with the ordered set of the possible solution of the required design. In the case of multi-parameter optimization, as well as in the case of vacuum mechanisms, the ordered set of the possible mechanism designs should be presented in the form of a multi-dimensional matrix. From one side, for this method of generation it is necessary to have a formal description of the designed mechanism parameters and, from another side, it is necessary to have the ordered array of different solution variables. The comparison of the required parameters with the ordered set of the solution variables makes it possible to formalize the process of selecting the variable.

Let us consider that the structure of every mechanism has a number K $(K \geq 2)$ of formal characteristics:

$$X_i = \{X_1; X_2; \ldots X_i \ldots; X_k\}, \tag{4.1}$$

and an array of properties and parameters of the object form a rectangular matrix:

E.A. Deulin et al., *Mechanics and Physics of Precise Vacuum Mechanisms*,
Fluid Mechanics and Its Applications 91, DOI 10.1007/978-90-481-2520-3_4,

$$\begin{array}{l} X_{11};\, X_{21};\, \ldots X_{i1};\, \ldots X_{k1} \\ X_{12};\, X_{22};\, \ldots X_{i2};\, \ldots X_{k2} \\ \vdots \\ X_{1j};\, X_{2j};\, \ldots X_{ij};\, \ldots X_{kj} \\ \vdots \\ X_{1m};\, X_{2m};\, \ldots X_{im};\, \ldots X_{km} \end{array} \tag{4.2}$$

where X_{ij} is a fixed link "property–structure" for the analyzed object, index i indicates the current number of the property or parameter being analyzed ($i = 1 \ldots k$), and index j indicates the current number of the structure type or parameter being considered ($j = 1 \ldots m$).

If the object to be designed has a large number of parameters or properties to be analyzed, and there is a large number of possible mechanisms corresponding to their structures, which can be compared, the matrix must be multi-dimensional. Every two-dimensional space marked by index "l" ($l = 1 \ldots n$) is included in this general multi-dimensional matrix, which has n spaces. In this connection, every decision variable of the designed object has to be marked by the index of the corresponding space.

Below, we will consider the method of a design generation using a consecutive two-dimensional matrix analysis. The main advantages of this method are the obviousness and the ability to generate fundamentally new types of mechanisms.

4.1 The Stages of the Matrix Method of the Mechanisms Generation

1. *Collection* and systematization of the *information* about ideas, patents, existing designs of the mechanism being designed, as well as the properties, parameters and systematization in correspondence with Equation (4.1).

2. *Analysis of the requirements demands* (tolerances) to the designed mechanism. Development of the criteria, which can be used for the mentioned properties, parameters comparison with the requirements (tolerances). When the designer applies the tolerances (tolerances, restrictions) to the created matrix the field of the satisfactory solutions corresponding to the equipment demands can be obtained.

3. Creation of the designed mechanism formal "image", i.e. *formal description* of the mechanism as it will be designed using the properties, parameters, structures used in the corresponding matrix. The range of the considered mechanism design (or scheme) for the adopted criteria (properties, parameters) determines the place of the considered scheme (design) in the space of the matrix. So, for example, the "image" of the long-travel ultrahigh vacuum feedthrough contains the following requirements:

1. Length of the travel: $L = 0 \div 1$ m

2. Work pressure: $P \leq 10^{-10}$ Pa
   (Outgassing flow: $Q < 10^{-9}$ $m^3 \cdot Pa \cdot s^{-1}$)
3. Vacuum features: oilless
4. Axial load: $P_{oc} \geq 100$ N
5. Precision (error): $\delta\varphi \leq 30$ microns
6. Longevity: $N > 10^6$ cycles
7. Laboriousness: $T \geq 150$ hours
8. Compactness (load/dimension): $P' \geq 0.5$ $N \cdot mm^{-1}$

4. Creation of matrix- classifications. The matrix is an n-dimensional space, where n is a number of the parameters, properties, which arrange the formal description of the designed mechanism according to pos. 3. Unoccupied cells of the matrix show the possibility to create new mechanisms with the combination of parameters and properties corresponding to the indicated point.

In traditional design, these properties-parameters combinations are usually considered as not promising (from the position of common sense) or not popular to the designer. Usually the designer considers these properties-parameters combinations as incompatible with the chosen model for designing.

During the design process it is necessary to create consecutively a system of two-dimensional matrices [1] beginning from the highest level of the matrix which reflects the basic principles of the design and finishing with the matrix at the n-th level, which reflects the minor aspects of the design. Usually, the technologist formulates the required parameters of the equipment to be designed and then these requirements are transformed into the properties-parameters limitations of the mechanism by the designer. The level of the matrix depends on the corresponding level of the requirements. For precise vacuum mechanisms the following parameters should be considered:

the first level parameters:

1. The components of the outgassing rate from the elements of vacuum feedthrough – m;
2. The type of the sealing element (mechanical, capillary, hydrostatic, dynamic) – y;
3. The principle of the force transmission, e.g. through the hole in the wall, using flexible wall or magnetic or electric field) – n;
4. The principle of the energy transmission (thermal, piezo electrical, magnetostrictive, electro-magnetic, mechanical) – l;

the second level parameters:

5. Gear-ratio (socket coupling $u = 1$, reduction gear $u = c$, rotary-linear motion transformer $u = c/\varphi$, step-down gearing $u < 0$, drive up gearing $u > 0$) – i;
6. Driving gear ratio type (planetary reducer, harmonic gear, nut-screw couple, etc.) – s;
7. Type of a load of sealing elements (e.g., the sealing bellow loaded by the torque of the feedthrough or, vice versa, unloaded by using gears; the sealing bellow loaded by the pressure difference or unloaded using stops, etc.) – z;

the third level parameters:

8. The sealing element form (membrane, thin-wall cylinder, Bourdon tube, bellow, thin-wall tube of a harmonic drive, etc.) – f;
9. The type of the deformation of a sealing element, e.g. the shape of the deformed bellow axis – α;
10. The type of device transmitting the movement (radial ball bearing, thrust bearing, spherical bearing, eccentric shaft, eccentric bush, double eccentric bush, crankshaft, inclined bush, etc.) – k;

the fourth level parameter:

11. Kinematic error – δ;
12. Friction speed in kinematic pairs in vacuum – v;
13. The system rigidity – j;
14. The maximal transference in vacuum in comparative units in respect to the basic element, for example bellow length L ($l = f(L), l > L, l < L$) – λ.

The general list of the parameters, which were used by different authors during the process of vacuum mechanisms design can be found in Section 4.2. This list can be expanded in the process of new mechanisms design.

It is necessary to say that the term "level" is a relative notion and reflects the connection of the considered parameters with the quality of the unit (machine, installation) being designed.

So, the symbol of the fixed-link "property-structure" of vacuum mechanisms or feedthroughs limited group which was selected on all levels of the matrix using the described method will be written as:

$$X = \begin{bmatrix} m & y & n & l \\ i & s & z & 0 \\ f & \alpha & k & 0 \\ \delta & v & j & \lambda \end{bmatrix}. \tag{4.3}$$

As was pointed out above, the matrix can be expanded. In this connection the matrix index had to be extended, first of all with the following parameters: "ultimate pressure", "cost", "reliability", "compactness", etc. Some of these parameters are correlated with the above-mentioned parameters. For example: bellows "reliability" is correlated with "type of the deformation of a sealing element", bellows sealed feedthrough "reliability" is correlated with the "type of a load on sealing elements", etc.) and can be used in the created matrices as reserved parameters.

Let us consider the application of the matrix method for the design of the ultra-high vacuum feedthroughs.

So, the matrix shown in Figure 4.1 presents the highest level of systematization in vacuum feedthroughs. It contains three subsets in four dimension which systemize the vacuum feedthroughs on the outgassing rate components: (1) seal fixing principle, (2) force-transmitting principle, (3) energy transformation principle. The combination of corresponding "principle" with the type of the sealing element, e.g.

Parameter \ Structure			Seal fixing principle X_{my}				Force transmitting principle X_{m0n}				Energy tranformation principle X_{m00l}				
m	Outgassing flow X_m	Type compressor	mechanic X_{m1}	capillary X_{m2}	hydrostatic X_{m3}	dynamic X_{m4}	solid shaft X_{m01}	flexible wall X_{m02}	magnet X_{m03}	electro-static X_{m04}	thermal X_{m001}	electro-magnetic X_{m002}	magneto-striction X_{m003}	piezo-electric X_{m004}	mechanic X_{m005}
1	$Q=q_e F_e+q_1 F_1+q_d F_2+Q_{i\,Lg}$	Oil seal	X_{11}	X_{12}			X_{101}								X_{1005}
2	$Q=q_d F_d+q_2 F_1+Q_{i\,Lg}$	Oilless seal	X_{21}				X_{201}	X_{202}	X_{203}						X_{2005}
3	$Q=q_c F_c+q_1 F_1+q_d F_2+\frac{kq'Fp^{1/j}}{\delta}$	Liquid seal		X_{32}	X_{33}	X_{34}	X_{301}							X_{3004}	X_{3005}
4	$Q=q_1 F_1+Q_{Lg}$	Gaseous seal	X_{41}			X_{44}	X_{401}								X_{4005}
5	$Q=\frac{q''Fp^{1/j}}{\delta}+q_i F_i$	Thin walled seal	X_{51}					X_{502}	X_{503}	X_{504}			X_{5003}	X_{5004}	X_{5005}
6	$Q=q_i F_i$	Autonomous drive						X_{602}	X_{603}	X_{604}	X_{6001}	X_{6002}		X_{6004}	

Fig. 4.1 Classification matrix of the first (highest) level of hierarchy for ultrahigh vacuum mechanical motion feedthroughs.

Wilson seal, liquid seal, gas seal, all-metal flexible seal, determines the number of the subsets of the lower level in every subset. In this case this level characterized as the structure of the outgassing rate also determines the ultimate pressure in the vacuum system.

Different classifications can be used for development of these matrices in a traditional style in which the objects are systemized only by one of the parameters. So, for example, the feedthrough classifications presented in [2, 3] were used as a basis for the matrix presented in Figure 4.1.

The consecutive improvement of new criteria of the lower level gives a possibility to present the analyzed sets as matrices of the lower level correspondingly. So, for example, the fixed link "property-structure" X_{502}, Figure 4.1, can be presented as a three-dimensional matrix, Figure 4.2, and a number of two-dimensional matrices presented in Figures 4.3, 4.4, 4.5, 4.6, 4.7.

The matrix presented in Figure 4.2 shows the realized combinations of fixed links "property-structure" of a linear-motion transmitting device, which transmits the motion through the solid wall using a flexible wall. This matrix consists of two two-dimensional matrices and shows the realized structures of the linear-motion ultrahigh vacuum feedthroughs based on oscillating and harmonic drives. The fixed links are presented as combinations of the parameters "relative transference" and "friction speed in vacuum". The last one is correlated with the "reliability" parameter $P(T)$.

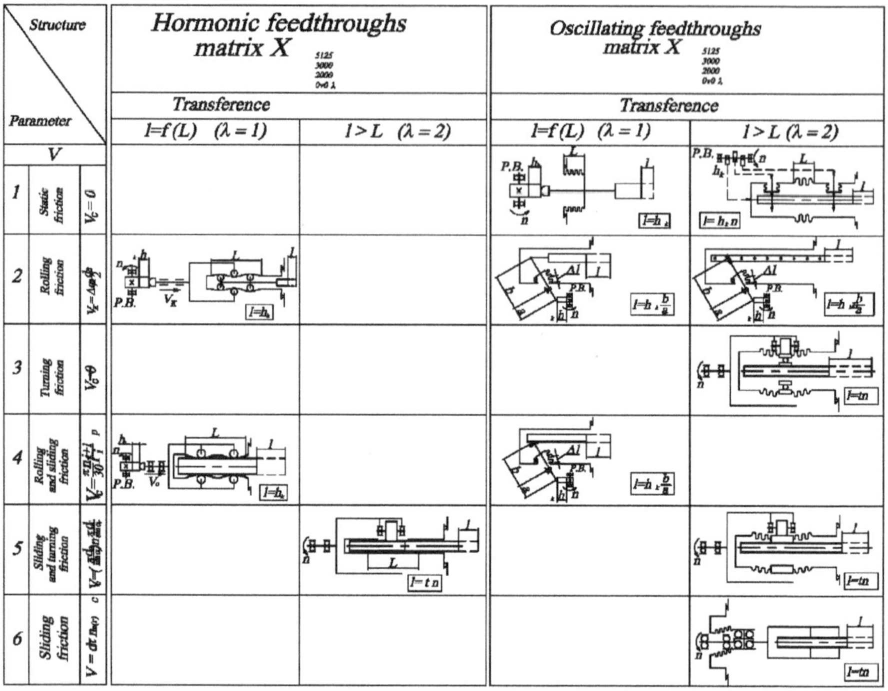

Fig. 4.2 Classification matrix at ultrahigh vacuum motion feedthroughs through a flexible wall.

The matrix presented in Figure 4.3 shows the realized principles of rotary- and linear-motion transmission with different transmission ratios in combination with different bellows eccentricity (fixed link "structure-shape of deformed bellows axis"). The matrix presented in Figure 4.4 shows the variables of bellow-sealed rotary-motion transmission in combination with different bellows eccentricity (fixed link "bearing structure-shape of deformed bellow axis").

The matrix presented in Figure 4.5 shows the realized principles of rotary-motion transmission with different structure of transmission ratio [4] in combination with different bellows eccentricity (fixed link "gear ratio-shape of deformed bellows axis").

All matrices presented in Figures 4.3, 4.4, 4.5 include the same parameter "shape of the deformed bellow axis" and all of them can be considered as one four-dimensional matrix.

If we consider the right part of the matrix presented in Figure 4.3 (linear-motion feedthroughs, $s = 4$) and if we use one additional parameter "friction speed" we come to another matrix, which is presented in Figure 4.6. The use of this matrix for the design of the ultrahigh vacuum linear-motion feedthroughs, helps us to find new solutions and to generate a number of new feedthroughs described in Chapter 8.

5. The selection of the basic scheme. This stage is a result of the comparison of the required parameters' properties with the set of the fixed links "property-

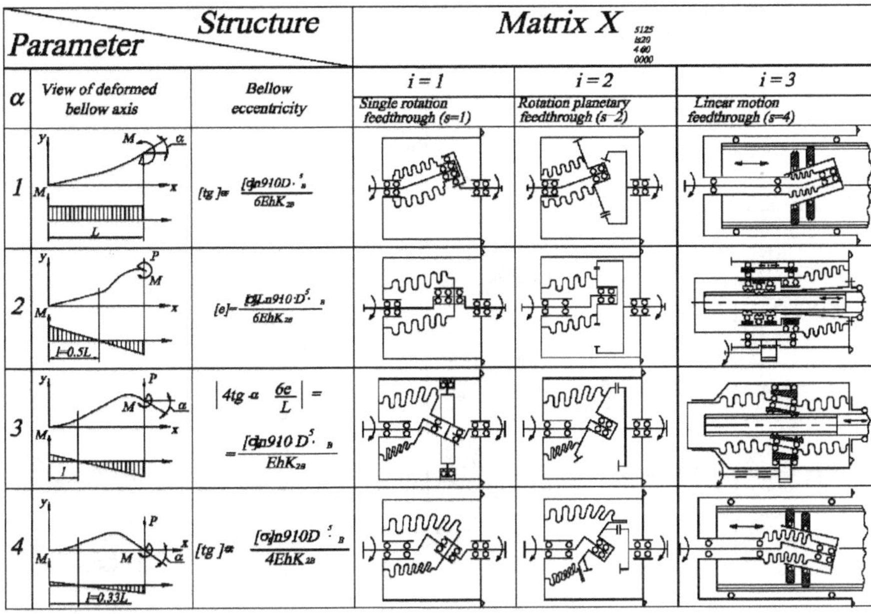

Fig. 4.3 Classification matrix for ultrahigh vacuum motion feedthroughs with bellows sealing.

structure" systemized in a matrix. The result of this comparison shows that a part of a matrix contains a limited set of possible solutions (designs). The task of the designer is to find the optimal design in this limited set. So, during the process of the design of standard ultrahigh vacuum rotary-motion feedthroughs with bellows sealing [5], the fifth line of the matrix (X_5 set) of the highest level, Figure 4.1, was chosen according to the requirements to limit the outgassing rate. According to the requirements of reliability, the first and the second lines of the second level matrices, Figures 4.3 and 4.4, were chosen because only these forms of the bellows deformation ensure the required longevity and reliability. In the continuation of the design process the two left columns of the matrix shown in Figure 4.4 were chosen, because only these types of bearings ensure the required reliability in ultrahigh vacuum.

The last four schemes of the chosen set situated in the left upper corner of Figure 4.4 were compared one with another on "laboriousness" criteria, therefore a result two basic schemes were chosen:

$$X = \begin{bmatrix} 5 & 1 & 2 & 5 \\ 1 & 1 & 0 & 0 \\ 4 & 3 & 1 & 0 \\ 0 & 0 & 0 & 0 \end{bmatrix} \quad \text{and} \quad X = \begin{bmatrix} 5 & 1 & 2 & 5 \\ 1 & 1 & 0 & 0 \\ 4 & 2 & 2 & 0 \\ 0 & 0 & 0 & 0 \end{bmatrix}. \tag{4.4}$$

The zeros in the indices are used for the parameters which were not used in the searching (designing) process.

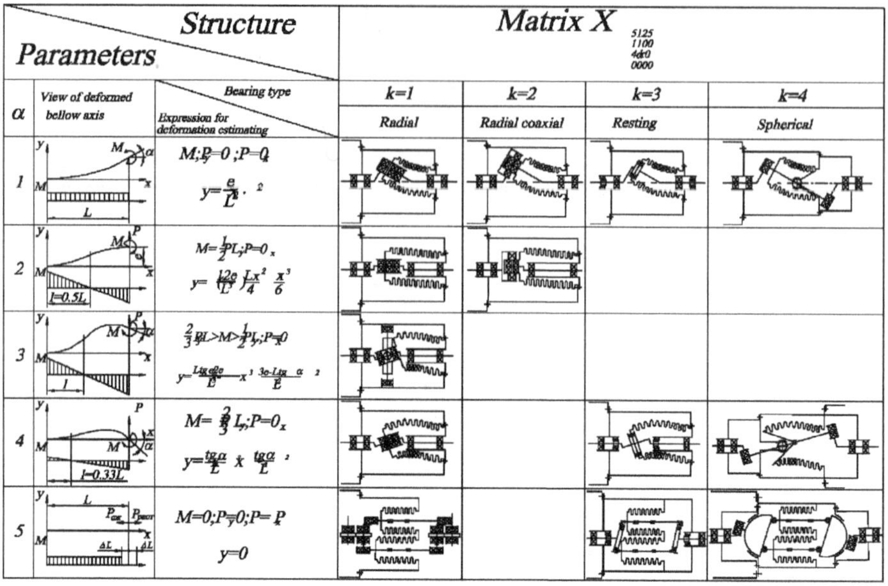

Fig. 4.4 Classification matrix for ultrahigh vacuum rotary-motion feedthroughs with bellows sealing.

The chosen schemes were used for development of a National Standard of bellows sealed rotary-motion feedthroughs in Russia [6].

For the design of ultrahigh vacuum long-travel linear-motion feedthroughs for use in the systems for the assembling of photoelectron gauge in vacuum (see Figure 1.9, on the highest level of the matrix), the same set (the fifth line of the matrix (X_5 set)) was chosen, but finally, on the second level of the matrix a different scheme was chosen. The index of this scheme is:

$$X = \begin{bmatrix} 5 & 1 & 2 & 5 \\ 1 & 1 & 0 & 0 \\ 4 & 2 & 1 & 0 \\ 0 & 0 & 0 & 0 \end{bmatrix}. \tag{4.5}$$

The bellow-sealed ultrahigh vacuum long-travel feedthrough of non-coaxial nut-screw type corresponding to this scheme [7] is shown in Figure 2.5. This design was also adopted as a National Standard.

6. The new basic scheme generation. The systemized structure of a matrix gives a possibility to put the attention of the designer on the "empty cells" of a matrix situated in the potentially prospective areas of positive solutions. It means that the matrix method (in contrast to the traditional classification methods), gives a real instrument to a designer for a new mechanism generation, because it shows the

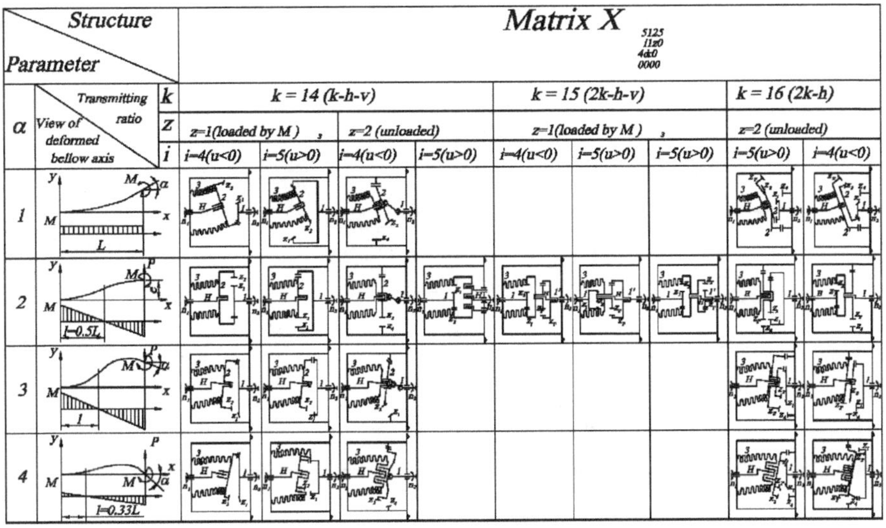

Fig. 4.5 Classification matrix for ultrahigh vacuum planetary gear rotary-motion feedthroughs with bellows sealing.

combination of properties-parameters principles, which can be used in the designing process to ensure the required parameters of the mechanism being created.

This method presents a standard procedure of the invention generation, which was successfully used for the new designs of linear-motion feedthrough [8–11]. Also as it was successively used for the new design of multi-coordinate magnetic rheology drive presented in Chapter 6.

4.2 The List of the Parameters of Vacuum Mechanisms Which Are Used in Matrix Analysis

The expert's assessment on the role of the parameters of the highest level can help to reduce the considered multitude (set) of the schemes (designs). The second level parameters and others should be reduced as much as possible. Below we consider the parameters of different levels which were used for the design of precise vacuum mechanisms.

4.2.1 The First (Highest) Level Parameters

- The structure of the outgassing rate from the vacuum drive:

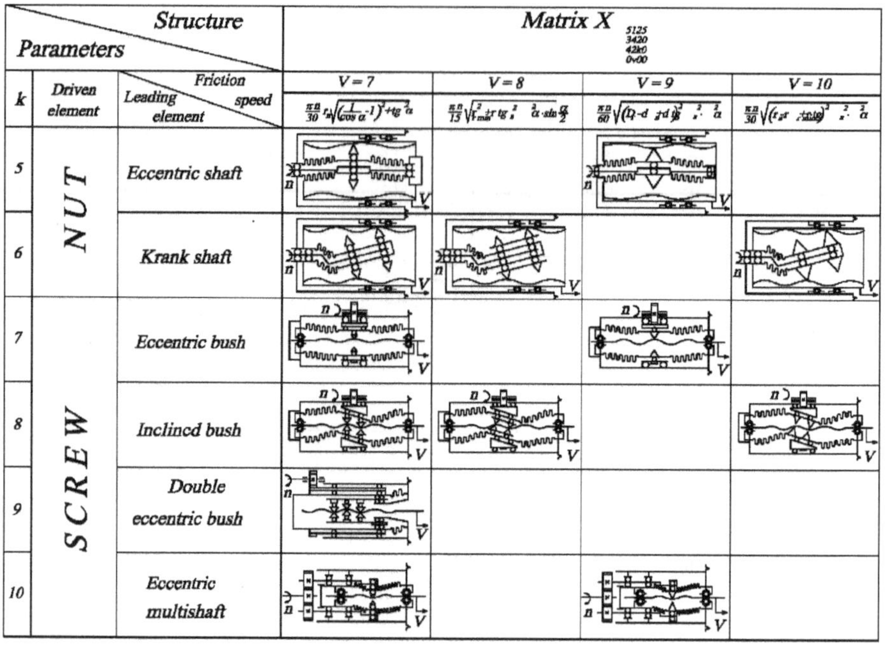

Fig. 4.6 Classification matrix for ultrahigh vacuum linear-motion feedthroughs with bellows sealing.

$$
\begin{aligned}
m &= 1 - Q = q_e F_1 + q_s F_2 + \Sigma q_i F_i + Q_l;\\
m &= 2 - Q = q_s F_2 + \Sigma q_i F_i + Q_l;\\
m &= 3 - Q = q_e F_1 + q_s F_2 + \frac{k q' F_2 p^{1/j}}{\delta};\\
m &= 4 - Q = \Sigma q_i F_i + Q_l;\\
m &= 5 - Q = \frac{q'' F p^{1/j}}{\delta} + \Sigma q_i F_i;\\
m &= 6 - Q = \Sigma q_i F_i,
\end{aligned}
$$

where Q is the total rate, m^3·Pa·s^{-1}; Q_l is the leakage, m^3·Pa·s^{-1}; q_e is the specific flow of evaporation, m^3·Pa·s^{-1}·m^{-2}; q_s is the specific outgassing flow of sealing materials, m^3·Pa·s^{-1}·m^{-2}; q_i is the specific outgassing flow of unbaked elements, m^3·Pa·s^{-1}·m^{-2}; q' is the specific flow of the gases penetrating through the liquid, m^3·Pa·s^{-1}·m^{-2}; q'' is the specific flow of the gases penetrating through the thin walls, m^3·Pa·s^{-1}·m^{-2}; F_1 is the area of the surface covered with the lubricant, m^2; F_2 is the area of the elastomer surface faced in vacuum, m^2; F_3 is the area of the thin-wall sealing elements, m^2; F_i is the area of the unheated elements faced to vacuum, m^2; k is the mixing coefficient of the liquid; δ

is the thickness of the thin-wall sealing element, m; p is the pressure difference on the sealing element, Pa.

- Seal fixing principle:

$y = 1$ – mechanical;

$y = 2$ – capillary;

$y = 3$ – hydrostatic;

$y = 4$ – dynamic.

- Force transmitting principle:

$n = 1$ – by a solid shaft;

$n = 2$ – through a flexible wall;

$n = 3$ – using a magnetic field;

$n = 4$ – using an electric field..

- Energy transformation principle:

$l = 1$ – thermal;

$l = 2$ – piezoelectric;

$l = 3$ – magnetostrictive;

$l = 4$ – electromagnetic;

$l = 5$ – mechanical.

4.2.2 The Second Level Parameters

- Gear-ratio:

$i = 1$ – coupling $u = 1$;

$i = 2$ – reducer $u = \text{const}$;

$i = 3$ – transformer of the rotation into linear movement $u = \text{const}/\varphi$;

$i = 4$ – underdrive $u < 0$;

$i = 5$ – drive-up $u > 0$.

- Type of transmitting mechanism:

$s = 1$ – bellows-sealed solid shaft;

$s = 2$ – planetary reducer;

$s = 3$ – harmonic gear;

$s = 4$ – nut-screw couple.

- Load on the sealing element (torque):

$z = 1$ – loaded;

$z = 2$ – unloaded.

4.2.3 The Third Level Parameters

- Type of flexible sealing element:

$f = 1$ – membrane;

$f = 2$ – cylinder tube;

$f = 3$ – Bourdon tube;

$f = 4$ – bellows;

$f = 5$ – thin-wall sealing gear.

- Shape of the deformed bellows axis: $\alpha = 1 \div 5$ – the shapes corresponding to this coefficient values are shown in Figures 4.3, 4.4, 4.5.
- Type of transmitting element:

$k = 1$ – radial ball bearing;

$k = 2$ – radial coaxial ball bearing;

$k = 3$ – thrust bearing;

$k = 4$ – spherical bearing;

$k = 5$ – eccentric shaft;

$k = 6$ – crankshaft;

$k = 7$ – eccentric bush;

$k = 8$ – inclined bush;

$k = 9$ – double eccentric bush;

$k = 10$ – set of eccentric shafts;

$k = 11$ – cylinder thin-wall sealing gear;

$k = 12$ – conical thin-wall sealing gear;

$k = 13$ – flat thin-wall sealing gear;

$k = 14$ – $k - h - v$ – planetary drive consisting of elements: sun-gear k, cage h, branch v [4];

$k = 15$ – $2k - h - v$ – planetary drive consisting of elements: two sun-gears k, branch v and imaginary cage h' [4];

$k = 16$ – $2k - h$ – planetary drive consisting of elements: two sun-gears k and cage h [4].

4.2.4 The Fourth Level Parameters

- Friction speed in kinematic pairs in vacuum:

$$\begin{aligned}
v &= 1 - V = 0;\\
v &= 2 - V = \Delta\varphi \frac{d_B}{2} n;\\
v &= 3 - V \to 0;\\
v &= 4 - V = f(n, h, m, b);\\
v &= 5 - V = \left(\frac{\pi d_B}{\sin\alpha_B} - \frac{\pi d_\Gamma}{\sin\alpha_\Gamma}\right) n;\\
v &= 6 - V = \pi d_{B(\Gamma)} n;\\
v &= 7 - V = \frac{\pi n}{30} r_B \sqrt{\left(\frac{1}{\cos\alpha} - 1\right)^2 + \tan^2\alpha};\\
v &= 8 - V = \frac{\pi n}{15} r_B \sqrt{r^2_{r\,\min} + r^2_{r\,\min}\tan^2\alpha \cdot \sin\frac{\alpha}{2}};\\
v &= 9 - V = \frac{\pi n}{60} \sqrt{(D_r - d_B)^2 + d_B^2 \tan^2\alpha};\\
v &= 10 - V = \frac{\pi n}{30} \sqrt{(r_B - r_{\Gamma\,\min})^2 + r_B^2 \tan^2\alpha},
\end{aligned}$$

where φ is the angular coordinate, deg; n is the speed of rotation, rev/min; h is the cam height, m; m is the gear teeth module, mm; b is the lever (leading gear radius equivalent), m; d is the mean diameter of the screw thread, m; d_Γ is the mean diameter of the nut thread, m; α_B is the screw lead angle, deg; α_Γ is the nut lead angle, deg; D_Γ is the mean diameter of the circular graves of the nut, m; r_B is the mean diameter of the screw thread, m; $r_{\Gamma\,\min}$ is the minimum mean diameter of the screw thread.

- The system rigidity:

$$\begin{aligned}
j &= 1 \quad \text{– flat contacting;}\\
j &= 2 \quad \text{– cylinder contacting;}\\
j &= 3 \quad \text{– spherical contacting;}\\
j &= 4 \quad \text{– disk spring.}
\end{aligned}$$

- Relative transference:

$$\begin{aligned}
\lambda &= 1 \quad l = f(L),\\
\lambda &= 2 \quad l > L;\\
\lambda &= 3 \quad l < L,
\end{aligned}$$

where l – resulting transference; L – length of the deformed sealing element.

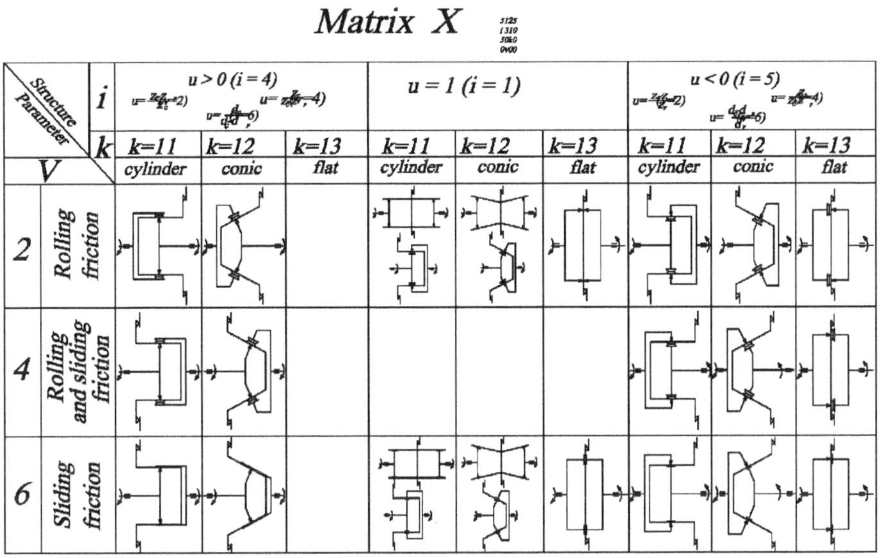

Fig. 4.7 Classification matrix for hermetic ultrahigh vacuum harmonic rotary-motion feedthroughs

$$X = \begin{bmatrix} 5 & 1 & 2 & 5 \\ i & 3 & 1 & 0 \\ 5 & 0 & k & 0 \\ 0 & v & 0 & 0 \end{bmatrix}.$$

The properties and parameters listed above have not been considered as unaltered constants, but rather as an interchangeable instrument for the ideas, schemes, decisions, analysis. If necessary this instrument can be replaced by another one or supplemented with new factors.

4.3 Algorithm of the Matrix Method of the Generation of New Mechanisms

The procedure of the scheme selection or generation, which is a basis for the design, can be described by the algorithm shown in Figure 4.8. The procedure called "Patent Library" realizes the input of the data, the "Designer" analyses these data and creates the formalized "images" of the mechanisms using the adopted criteria. At the same time, the "Technologist" formulates the requirements to the equipment for the mechanism being designed. By doing so on every level of the design (class equipment, group equipment, type equipment) a design draft becomes more detailed.

The input of the "Technologist" requirements compared with the parameters and properties of the analyzed mechanisms (comparison with the formal "images" of the mechanisms) gives the possibility to use both the unified marks for parameters

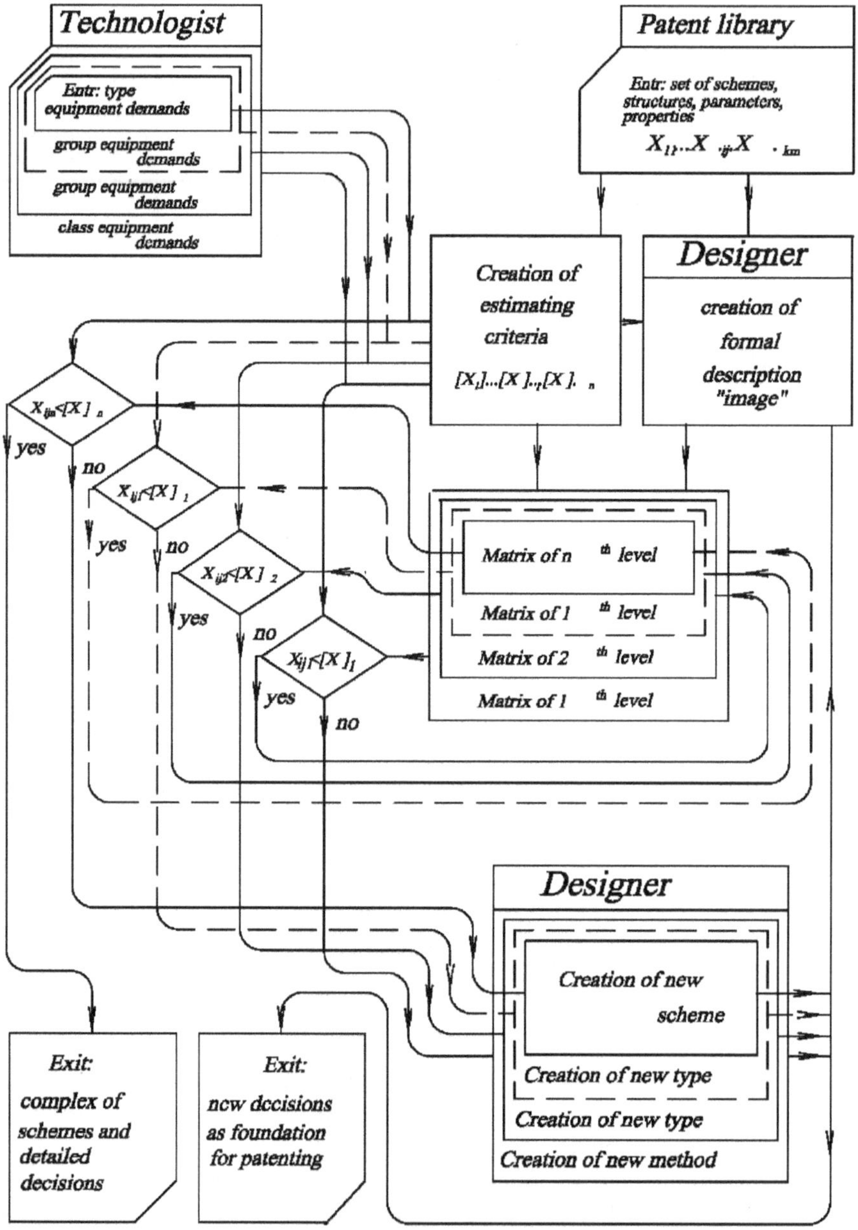

Fig. 4.8 Algorithm of the mechanism scheme selection.

limitations (tolerances) and the estimation of every scheme and every variable of the mechanism design to be placed in the matrix.

In the described algorithm, instead of one multi-space matrix, the system of two-dimension matrices of different levels is used. It gives the possibility to simplify the comparison procedure in each matrix and to do this procedure repeatedly in every dimension passing from one level to another.

The comparison procedure was done in the hierarchical matrices system by selecting the areas of positive solutions in each matrix. If the marked out area of the matrix does not contain any solution (if this area contains only "empty cells"), it means that the designer can find, generate or invent a new mechanism with the required parameters for the selected "property-parameter" combination. The invented mechanism had to be placed into the "empty bond" of the matrix.

After the procedure is finished in the matrix of the highest level, the searching procedure passes on to the matrices of the lower levels. The search is concluded in a satisfactory way, when within the limited set of schemes, the schemes which can be practically realized are chosen. So, during the design of the standard ultrahigh vacuum linear-motion feedthrough the finally limited set contains only two basic schemes:

$$X = \begin{bmatrix} 5 & 1 & 2 & 5 \\ 3 & 4 & 2 & 0 \\ 4 & 2 & 5 & 0 \\ 0 & 9 & 0 & 0 \end{bmatrix} \quad \text{and} \quad X = \begin{bmatrix} 5 & 1 & 2 & 5 \\ 3 & 4 & 2 & 0 \\ 4 & 2 & 7 & 0 \\ 0 & 9 & 0 & 0 \end{bmatrix}.$$

These schemes correspond to the column marked by symbol $V = 9$ and to the lines marked by symbols $k = 5$ and $k = 7$, Figure 4.6.

During the design of the ultrahigh vacuum planetary rotary-motion feedthrough, the finally limited "set" contains only one basic scheme:

$$X = \begin{bmatrix} 5 & 1 & 2 & 5 \\ 5 & 1 & 1 & 0 \\ 4 & 2 & 14 & 0 \\ 0 & 0 & 0 & 0 \end{bmatrix}.$$

This scheme corresponds to the column marked by symbols $k = 14$, $z = 1$, $i = 5$ and to the line marked by symbols $a = 3$, Figure 4.5.

During the design of the ultrahigh vacuum harmonic rotary-motion feedthrough, the finally limited "set" also contains only one basic scheme:

$$X = \begin{bmatrix} 5 & 1 & 2 & 5 \\ 4 & 3 & 1 & 0 \\ 5 & 0 & 11 & 0 \\ 0 & 4 & 0 & 0 \end{bmatrix}.$$

This scheme corresponds to the column marked by symbols $i = 4$, $k = 11$ and to the line marked by symbol $V = 4$, Figure 4.7. It is necessary to remember that the initial set of the designs contains more than 1200 schemes!

Now, the designer can use the matrices presented above as well as the newly created using the described principles and methods. The routine design of new schemes now is based on deeper physical analysis of potential abilities of the selected constructions or new designs.

Returning back to Figure 4.1 where the most complete set of schemes of vacuum feedthroughs is presented, we can see that the basic parameter of a vacuum mechanism, especially of an ultrahigh vacuum mechanism, is the outgassing rate, which limits the ultimate pressure of vacuum equipment.

At the same time for space mechanisms or for the mechanisms of nuclear equipment, the key parameters are longevity and reliability, which depend on the seal fixing principle, seal type, transmitting element shape. The presented method has ensured the creation of some types of standard vacuum feedthroughs: the collar feedthroughs [12], UHV Rotary-Motion Feedthroughs [6], Harmonic Drive Feedthroughs [13], UHV long travel linear-motion feedthroughs based on the non-coaxial nut-screw couple [7].

References

1. Deulin E.A., Method of selection of reliable vacuum mechanisms structure, in *Proceedings IVC-12*, 12–16 October 1992, Book of Abstracts, the Hague, 1992, p. 430.
2. Deulin E.A. et al., Unification of vacuum mechanical motion feedthroughs, in *Special Technological Equipment for Microelectronic Industry*, No. 3/29, 1971, pp. 77–85.
3. Roth, *Vacuum Sealing Techniques*, Pergamon Press, 1966.
4. Kudriavtsev V.N., *Planetary Gear Boxes*, Machinostroenie, Moscow, 1966, 307 pp.
5. Blinov I.G., Boreyko A.I., Uniform UHV rotary-motion feedthroughs, *Instr. Techn. Experiment*, No. 1, 1974, 168–170 [in Russian].
6. Deulin E.A., Russian Standard “UHV Rotary-Motion Feedthroughs”, Electrostandart, Moscow, 1973, 9 p.
7. Bushenin D.V., Boreyko A.I., Uniform UHV linear motion feedthroughs, *Instr. Techn. Experiment*, No. 1, 1983, p. 224 [in Russian].
8. Bushenin D.V., Deulin E.A., Marusov V.A. et al., Device for transmission of linear motion into vacuum, Soviet Patent No. 663938, 1979.
9. Bushenin D.V., Deulin E.A., Marusov V.A. et al., Device for transmission of linear motion into vacuum, Soviet Patent No. 681279, 1979.
10. Deulin E.A., Marusov V.A. et al., Device for transmission of linear motion into vacuum, Soviet Patent No. 889980, 1981.
11. Deulin E.A., Marusov V.A. et al., Device for transmission of linear motion into ultrahigh vacuum, Soviet Patent No. 8942991, 1981.
12. Deulin E.A., Russian Standard “Collar Feedthroughs”, Electrostandart, Moscow, 1979, 19 pp.
13. Deulin E.A., Russian Standard “Harmonic Drive Feedthroughs”, Technical Parameters, Electrostandart, Moscow, 1977, 7 pp.

Chapter 5
Precision of Vacuum Mechanisms

5.1 The Constituents of Errors of Vacuum Mechanisms

The required precision of vacuum installations in most technologies (thin film coating, electron beam welding, diffusion welding) is the same as in traditional engineering industry, but operating conditions of the mechanisms work are much worse, because of vacuum and periodical outgassing baking (see Tables 1.1, 1.2, 1.4). At the same time, the equipment of surface research, electron beam-, ion beam lithography, Table 1.3, atomic force scanning microscopy, Table 1.6, requires much higher precision of the mechanisms, which should work in ultrahigh vacuum (up to 10^{-9} Pa).

The first peculiarity of vacuum mechanisms is the existence of additional elements in their kinematic chains including: motion feedthroughs, sealing elements, devices for baking, heating elements, holes and grooves on the coupling surfaces for the evacuation of residual gases. All these elements reduce the rigidity of the drives and generate additional disturbing forces that result in the appearance of additional components in the total error. In some cases these new error components can be 10–100 times higher (!) than the errors of basic elements of transmission.

The second peculiarity of vacuum mechanisms is, from one side, the absence of a work load as that is contrast to atmospheric machines and, from another side, the existence of large and non-constant friction forces, additional forces generated by atmospheric pressure, forces generated by deformed sealing elements. It is necessary to emphasize the accidental character of friction forces and their non-linear character as well as anisotropy of elastic sealing elements and intensive wear of friction surfaces.

Let us compare the diagrams of summarized error components of the digitally controlled lathe support [1] and of the digitally controlled sample holder of the ultrahigh vacuum manipulator, of the molecular epitaxy installation [35], see Figure 5.1.

Comparison of the diagrams shows the dominating role of the elastic components of error for vacuum mechanism, that is evidence of the direct and indirect influence of vacuum, because the technological loads are absent in this case.

E.A. Deulin et al., *Mechanics and Physics of Precise Vacuum Mechanisms*, Fluid Mechanics and Its Applications 91, DOI 10.1007/978-90-481-2520-3_5,

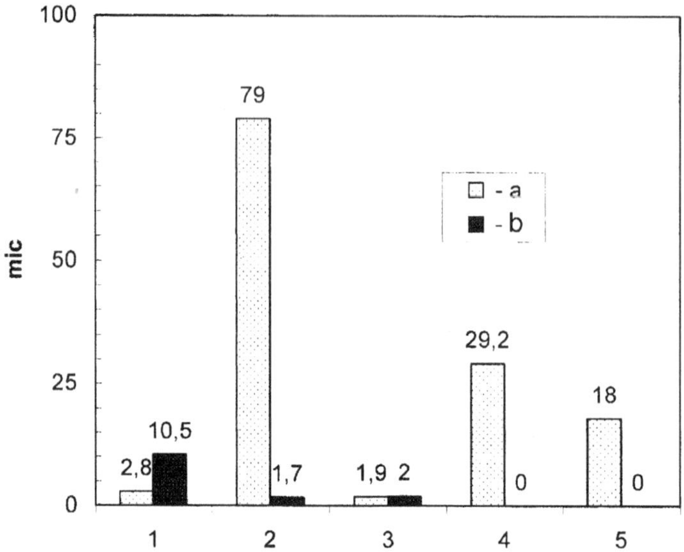

Fig. 5.1 The diagrams of total error components, mic: (a) ultrahigh vacuum manipulator; (b) digitally controlled lathe support; (1) kinematic error δ_1; (2) elastic error $\delta_5 + \delta_6 + \delta_7 + \delta_8$; (3) error of initial action and instrumental error $\delta_n + \delta_i$; (4) error of calculated action δ_v; (5) error of dead stroke and elastic error as a result of non-linear elastic forces $\delta_3 + \delta_9$.

Table 5.1 contains the list of the main components of the total error δ and the references to the basic publications where these components are described.

The most famous publications on the theory of precision are dedicated to the determination of the kinematic error δ_1, which is considered to be a result of the manufacturing error. However, this approach does not lead to the error determination of ultrahigh vacuum mechanisms, whose components are usually manufactured with very high precision, but the total precision of the mechanisms usually is much less than a designer expectations.

The problem of precise ultrahigh vacuum mechanisms development leads to the development of advanced theory of precision. The methods of calculation of all components of the total error of vacuum mechanisms were developed.

The precision of a vacuum drive depends on the method of the drive control and on the specific direct and indirect vacuum influences, which become apparent in the following issues:

- The scheme (see Chapter 4) and the design of the mechanism depends on the required vacuum (see Chapter 1).
- The main part of the vacuum drive could not be realized without bearings and other kinematic pairs (gears, nut-screw pairs, claw coupling, guides) and without sealing elements.
- The baked elements of the vacuum drive are designed with the increased gaps in the coupling to avoid seizing at temperature variation.

Table 5.1 Components of error.

The summarized error components				Reference
Control system type	Title		Error component name	
	Total	correct		
Open control system	K	1	Kinematic error as result of drive manufacturing	6,8,10, 16,17,18,19, 27,28, 30,33
		2	Kinematic error as result of drive wear	19,24,26
	L	3	Dead centre of kinematic elements	16,17,18
		4	Dead centre as result of wear	24,26
	c	5	Elastic error as result of a load	29,31,32,33
		6	Elastic error as result of friction force and atmosphere pressure	26
		7	Elastic error as result of friction force at movement	34
		8	Elastic error as result of friction force variation	34
		9	Elastic error as result of non linear elastic forces	4
		10	Error of occasion kinematic component	16,17
	L	11	Error of occasion dead stroke	16
	c	12	Error of occasion rigidity	20, 31,32
		13	Error of occasion friction force.	34
		14	Error of variated friction forces	5
Loop controlled system	B		Error of calculated action	5,9,19
	H		Error of initial action	5,9,19
	C		Error of static resistant force	34
	L 3		Error of dead stroke	16,17,18
	I		Instrumental error	5,9,19

- The sealing elements (bellows, thin-wall sealing elements, cuffs) create an additional non-linear load to the drive.
- There is a variable load to the sealing elements of the drive as a result of the atmospheric pressure, which disappears after the atmosphere or work gas is let into a vacuum chamber.
- The set of the constructional materials, coating materials, lubricants is limited by the technological process limitations.
- The dimensions of a sealing element often limits the length of the travel of the drive.

Let us consider the total error of outlet elements of the drive. This error can be expressed as the difference between the expected φ_0 and the actual φ_D position of the outlet element:

$$\delta\varphi_D = \varphi_0 - \varphi_D = \delta_V + \delta_N + \delta_k + \delta_L + \delta_c + \delta_I, \tag{5.1}$$

where δ_V is the error component due to master control action; δ_N is the error component due to initial action or dither signal action; δ_k is the error component due to kinematic error of the drive; δ_D is the error component due to backlash of the drive; δ_c is the error component due to friction forces and another resistance forces; δ_I is the instrumental error component caused by the measuring error.

In the drives with the open-loop control (when the output element of the drive is not controlled by the sensor), the total error of the transference is the result of the errors of the different elements of the drive summarizing.

Examples of the open-loop-controlled system are the molecular epitaxy installations, installations of Auger analysis, Secondary Ion Mass Spectrometry and other systems for surface analysis. In the drives of these systems all components of the total error are present except an instrumental error component because of the absence the positioning sensors in this installation.

Thus, Figure 5.2a shows the drive scheme of the molecular beam epitaxy installation. The error components δ_V and δ_I are formed in the drive 1 at the moments of malfunction of the control system or the interruption of the drive supply; the error components δ_k and δ_L are the errors of the transmitting elements (rotary-motion bellows-sealed feedthrough 2 or nut-screw couple 3), error component δ_c is the error of bellows-sealed feedthrough 2 and nut-screw couple 3, which are the result of the force of atmospheric pressure on the bellow walls and also the result of anisotropy of elastic properties of the bellow and increased friction forces in vacuum. A structure diagram of this scheme is shown in Figure 5.3.

Partial or complete loop-controlling of the controlling system allow partially or completely exclude the number of the error components which are formed in the loop-controlled zone (marked *33*) using the positioning sensor and a signal of the loop control (signal φ_{oc}). However, add the error component of the used sensor δ_I.

Figure 5.2b shows the scheme of the manual drive 4 of the manipulator carousel of the installation of the surface analysis which has the loop-controlled zone φ_{oc} and where the visual checking of the limb 7 positioning of the handle 5 apply to exclude the error component δ_I. An error component δ_V depends on the operator's eye vision, and the error component δ_I depends on the precision of the limb or another counting device. The error components δ_k, δ_L, δ_c are formed in the nut-screw pair 3 and in the bellows-sealed feedthrough 6. The structure scheme of this drive is shown in Figure 5.3b.

Partial loop-controlling of the system is used in the drives of the welding electron beam installations, in the drives of the ion implantation installations, electron beam lithography installations and in vacuum transporting systems. So, for example, the special disk with the holes 8, Figure 5.2b, is used in combination with the photon-coupled device 9. This system excludes the error components φ_{oc} which appear in the loop-controlled zone (in the elements 1, 7, left gear coupling) but leave the errors δ_k, δ_L, δ_c which appear in the kinematic elements 3, 10, 7 (right gear coupling) in the open-loop-controlled zone (P3). The structure scheme of this drive is shown in Figure 5.3c.

The drives with complete loop-control of the controlling system allow to exclude almost completely the main number of the error components shown above. It is a

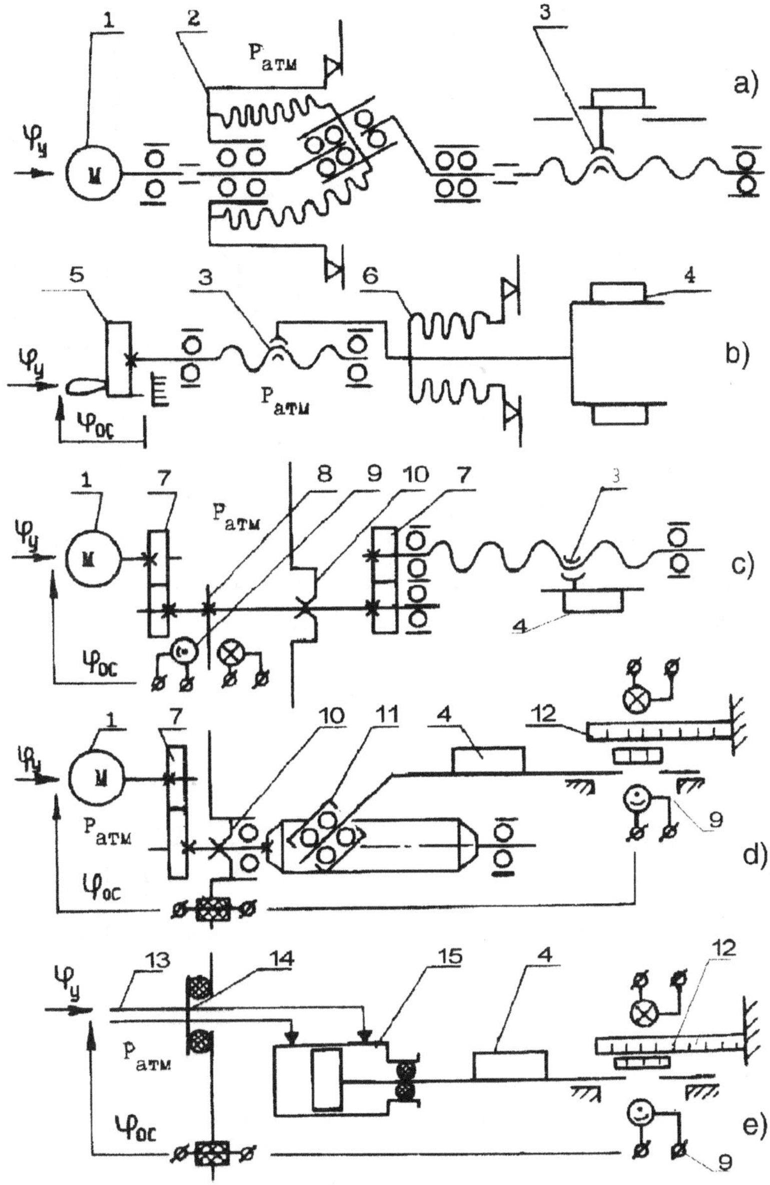

Fig. 5.2 Kinematic schemes of vacuum mechanisms: (a) the drive scheme of the molecular epitaxy installation; (b) the drive scheme of the manipulator carousel of the installation of the surface analysis; (c) the drive of the welding electron beam installations; (d) the completely loop-controlled drive of the electron beam installation; (e) the hydraulic drive of the electron beam lithography installation; 1 – the drive; 2 – rotary-motion bellows sealed feedthrough; 3 – nut-screw couple; 4 – the transferred sample; 5 – the limb; 6 – the bellows-sealed feedthrough; 7 – the kinematic elements; 8 – the special disk with the holes; 9 – the photon-coupled device; 10 – the sealed elements; 11 – the kinematic elements; 12 – the raster sensor; 13 – the hydraulic system; 14 – the vacuum chamber wall; 15 – the hydrocylinder (hydraulic drive).

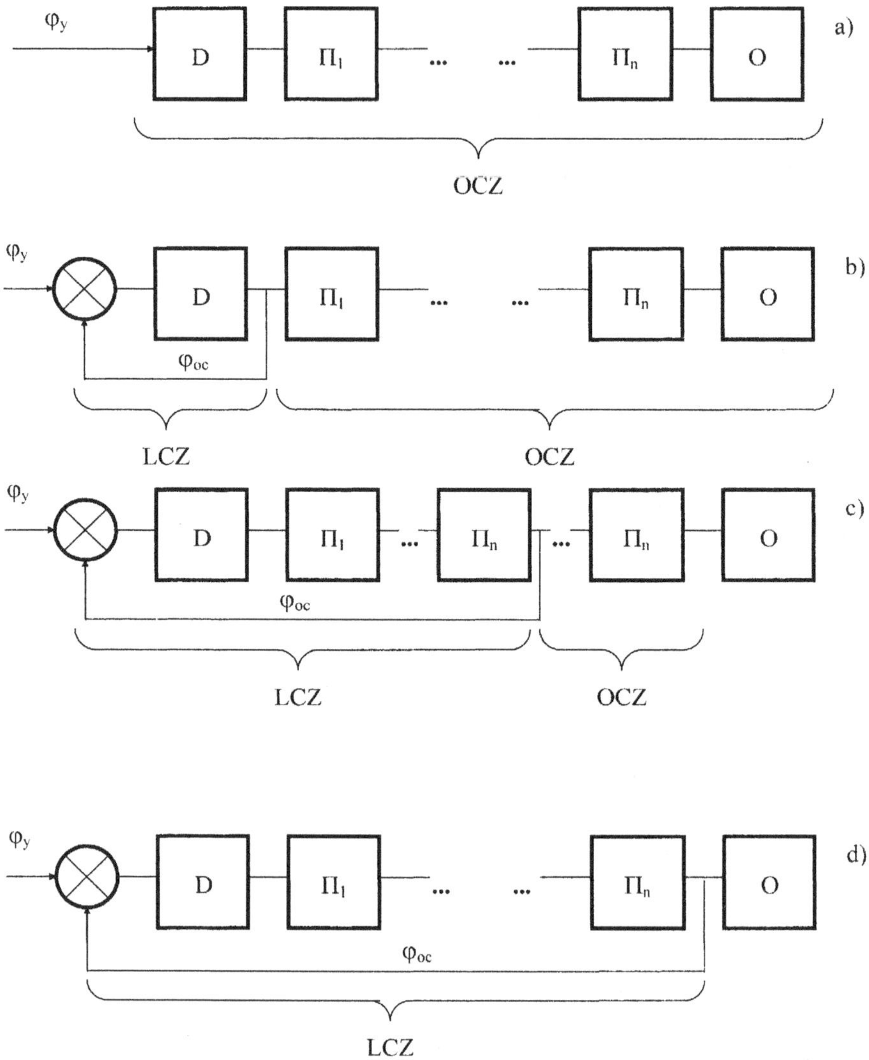

Fig. 5.3 Structure diagrams of the drives with different controlled zone (corresponding to Figure 5.2); OCZ – open controlled zone; LCZ – loop-controlled zone.

reason that such drives are used in those installations, where the precise motion is required, i.e., in the installations of electron-, ion lithography, electron microscopy, precise electron or ion processing and so on.

The example of such a drive is the completely loop-controlled drive of the electron beam installation shown in Figure 5.2d. The positioning sensor, i.e. elements of the raster photon-coupled system 12 is based on the same carrier with the transferred sample 4. Sometimes, a raster element can be used as the worked substrate (wafer) that allows to exclude the workpiece positioning error.

The error of movement of a completely loop-controlled drive of vacuum installations can be expressed:

$$\delta\varphi_D = \delta_N + \delta_L + \delta_c + \delta_I. \tag{5.2}$$

The total error of the installations of electron-, ion lithography in the correspondence with requirements from Chapter 1, should not be higher than 1 to 2 microns. The required precision of scanning microscopes is much higher and is about 0.05 microns.

The problem with a loop-controlled drive is the limitation of temperature due to gas release. There is an increased outgassing rate because of the necessity to fix the precise measuring devices (positioning sensors) into the vacuum chamber on the object being transferred.

The backlash and large friction forces in vacuum increase the dead zone of the drive (increase the δ_L and δ_c components of the total error). The designers of precise vacuum drives try to eliminate kinematic pairs from a vacuum chamber, to shorten kinematic chains, to replace an electromechanical drive by a hydraulic drive, to get an autonomous drive without the use of vacuum feedthroughs.

An example of such a decision is the autonomous hydraulic drive of the electron beam lithography installation. The scheme of this drive is shown in Figure 5.2e. The control signal in the form of a controlled oil flow goes inside the sealed pipe through the vacuum chamber wall 14 to the hydrocylinder (hydraulic drive) 15. The hydrocylinder moves the support with the wafer and the raster sensor 12a. The smallest leakage of a liquid from the hydrocylinder contaminates the residual atmosphere in the vacuum chamber. It compels to search other ways of improving the precision of vacuum drives.

The different ways of the precision increasing of vacuum drives are analyzed in the material presented below. We shall consider the principles of total error formation in the open-loop-controlled drives as well as in the completely loop-controlled drive.

5.2 The Basic Positions of the Precision Theory of Vacuum Mechanisms

5.2.1 Open-Loop-Controlled Drive

Let us consider the process of open-loop-controlled drive precision calculation.

1. The components δ_k, δ_L, δ_c of the total error in an open-loop-controlled drive are realized *in corpora* as the kinematic error of the drive δ_1 (see Table 5.1) or as the error of transference $\delta_{\varphi\Sigma}$ (see Figure 5.4), as the recoil error of the drive $\mathbf{j}_{\varphi\Sigma} =$

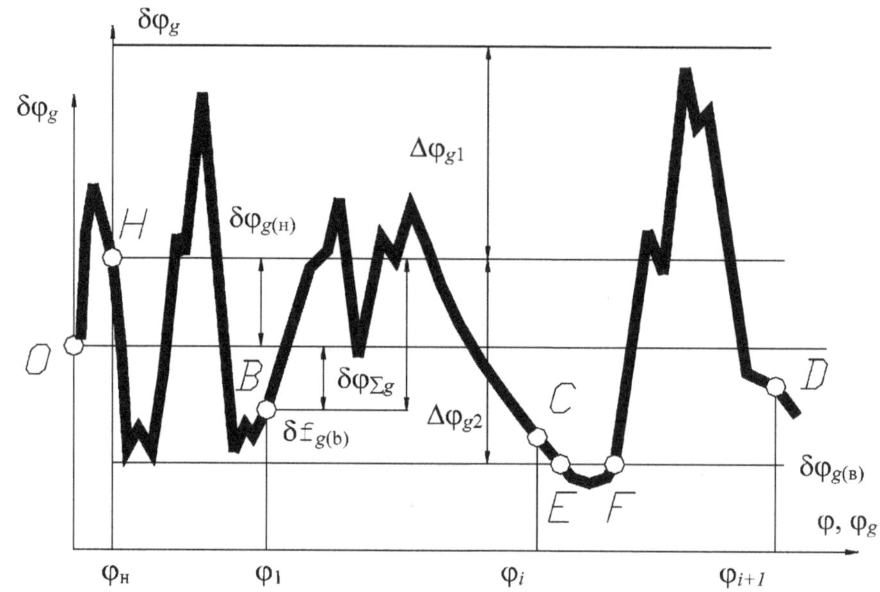

Fig. 5.4 Consideration of the terms "positioning error", "transference error". The symbols are: TO – the point of the account beginning the beginning of theoretically considered transference. TH – the point of the real transference beginning. $\varphi = \varphi_i, \varphi_0, \varphi_g$ – the transference of real, theoretical kinematic chains (USSR standard 21098-82) also as transference of real kinematic chain, respectively. $\delta\varphi$ – the error or real kinematic chain (USSR standard 21098-82). $\delta\varphi_{gb}$ – the real position error in point B. $\delta\varphi_{\Sigma gb} = \delta\varphi_{gH} - \delta\varphi_{gb}$ – the real transference error in the point B. $\Delta\varphi_{g1}$, $\Delta\varphi_{g2}$ – the top and the down error, respectively.

$\delta_3 + \delta_4 + \delta_{11}$ and as the elastic error $\delta)\varphi y = \delta_5 + \delta_6 + \delta_7 + \delta_8 + \delta_9 + \delta_{12} + \delta_{13} + \delta_{14}$, correspondingly, because of the absence of a movement sensor.

Usually, the value of "intrinsic error" of the drive without the components, which are caused by the external factors (i.e. without components δ_V, δ_N, δ_I, see Equation (5.1)) is of special interest. In this case the error of positioning can be written:

$$\delta\varphi_D = \varphi_D - \varphi_0 = \delta_k + \delta_L + \delta_c = \delta_1 + \mathbf{j}_{\varphi\Sigma} + \delta\varphi_y, \tag{5.3}$$

where δ_1 is the kinematic error of the drive; $\mathbf{j}_{\varphi\Sigma}$ is the total recoil error of the drive; $\delta_{\varphi y}$ is the total elastic error of the drive.

Let us consider how the error components of the drive have an influence upon the manufacturing process of the technological equipment.

The automated technological equipment which uses the precise drive can be divided into two groups according the principle of the spoilage appearance as result of the tolerance overrun by the total error of the drive.

The mechanisms of the first group of equipment work in the positioning mode of operation. The examples of this group of equipment are: Auger-control, microprobe, electron-beam lithography installations and ion-beam lithography installations. The initial reference point in this equipment usually is the random point φ_N on the axis φ of the transference (see Figure 5.4), which corresponds to the beginning of the technological process or to the reference mark on the object being transferred. In the last case the reference mark positioning is detected with the error $\delta_I = \delta\varphi_{gN}$ – error of measuring system. Let us apply that the abscissa (Y-coordinate) graph (Figure 5.4) determines an output error of the mechanism. In this case point, **O** corresponds to the position of the "ideal" theoretical kinematic chain according to the Russian Standard.[1] Correspondingly, if the point **H** corresponds to the initial position of the mechanism the point **H** abscissa $\delta\varphi_{gH}$ determines the initial output error of the mechanism in respect to the ideal "theoretic kinematic chain", while if the point **B** corresponds to the first point of the positioning of the mechanism, the point **B** abscissa $\delta\varphi_{gB}$ determines the first output error of the mechanism in respect to the ideal "theoretic kinematic chain". The true total error of positioning for point **B** can be determined as an algebraic difference of two above-mentioned errors:

$$\delta_{\varphi\Sigma} = \delta\varphi_{gB} - \delta\varphi_{gH}. \tag{5.4}$$

The error $\delta_{\varphi\Sigma}$, which is determined by Equation (5.4) is different from that determined by the Standard (see footnote 3), since it includes not only kinematic and recoil error components of the mechanisms and details but also elastic components of the error.

The parametrical reliability on "precision" parameters of positioning of a drive can be presented as a probability that the positioning error of the drive $\delta_{\varphi\Sigma}$, is smaller than the limit of the drive error $\Delta\phi_{gi}$ in the positioning points:

$$P(\delta\phi_g) = \Phi(\Delta\phi_{g1} > \delta_{\varphi\Sigma i} > \Delta\phi_{g2}), \tag{5.5}$$

where $\Delta\phi_{g1}$, $\Delta\phi_{g2}$ is the upper and lower limits of the drive error; $\delta_{\varphi\Sigma i}$ is the drive error in the ith point.

Since the values of all error limits are counted from the initial point of transference, inasmuch the errors being counted receive the additional component of the error $\delta\varphi_N$, which describes the position (the error) of the initial point in respect to the "theoretical kinematic chain". It is obvious, that all errors of positioning, as in the point of the initial positioning (point H) also as in the subsequent points (pp. B, C, D, Figure 5.4) are being described by the same function, but by different parts of it.

The true expected value of the total error of positioning in the ith point of positioning $M(\delta_{\varphi\Sigma i})$ can be written in much the same equation (5.4):

$$M(\delta f_{\Sigma i}) = M(\delta\varphi_{gi}) - M(\delta\varphi_{gH}), \tag{5.6}$$

---

[1] Russian Standard: 21098-82 (Kinematic chains. The methods of calculation of precision, Moscow, Standard edition, 1984, 26 pp. [in Russian].

where $M(\delta\varphi_{gi})$ is the expected value of the total error of positioning in the ith point of positioning. The standard deviation can be determined:

$$\sigma(\delta\varphi_{\Sigma gi}) = \sqrt{\left[\frac{\partial(\delta\varphi_{\Sigma gi})}{\partial(\delta\varphi_{gi})}\right]^2 \sigma^2(\delta\varphi_i) + \left[\frac{\partial(\delta\varphi_{\Sigma gi})}{\partial(\delta\varphi_{gH})}\right]^2 \sigma^2(\delta\varphi_{gH})}. \qquad (5.7)$$

In the most general case, when the standard deviation in the initial point and in the point of positioning are equal, and the error components δ_I, δ_V, δ_N, (5.1) can be neglected, the estimation of the standard deviation can be simplified:

$$\sigma(\delta\varphi_{\Sigma i}) \approx 1.4\sigma(\delta\varphi_{gi}). \qquad (5.8)$$

The equipment of the first group has satisfying results and good quality of the output at the positioning in the points with coordinates $\phi_1, \ldots, \phi_i, \phi_{i+1}$, if these points are situated within the tolerance band $[\Delta\varphi_{g1} - \Delta\varphi_g]$. This equipment ensures a good quality of the output (in "precision" parameter) even in the case when the error function is out of the tolerance band in the intermediate sites, for example, the site EF on Figure 5.4.

The second group of equipment works in the uninterrupted mode of operation and the waste appearance takes place in the case of exceeding the tolerance band in any point of the transference site. In this case the error function must be considered as an uninterrupted accidental function. The probability of the "parameter" breakdown on the "precision" parameter can be written:

$$\left.\begin{array}{l} P(\delta\varphi_g) = F(\Delta\varphi_{g1} > \delta\varphi_{\Sigma g} > \Delta\varphi_{g2} \\ \delta\varphi_\Sigma = f(\varphi_i - \varphi_u) \end{array}\right\}, \qquad (5.9)$$

where $(\varphi_i - \varphi_u)$ is the working site of the transference.

In the example shown in Figure 5.4 this group equipment ensures the good output only on the transference interval $\phi = \phi_i - \phi_H$. However, on the interval $\phi = \phi_{i+1} - \phi_i$ it ensures a waste output because of the error appearance out of the tolerance band on the site EF. The examples of this group of equipment are: crystal growing equipment, zone-melting equipment, electron-beam lithography installations and ion-beam lithography installations with uninterrupted movement of the table.

5.2.2 Completely Loop-Controlled Drive

Let us consider the precision of the loop-controlled drive of an electron beam installation. As it was shown above, the precision of a loop-controlled system (in our case the precision of positioning) depends on the error of master control action (δ_V) and the error of initial action or dither signal (δ_N) also as on the recoil of the drive (δ_D) and on instrumental error component (δ_I). So, the error of positioning of the loop-controlled drive can be presented in the following view:

$$\delta\varphi_\Sigma = \delta_V + \delta_N + \delta_{CT}, \tag{5.10}$$

where δ_V is the error of master control action (when using the stabilization system it can be allowed that $\delta_V = 0$); δ_N is the error of initial action or dither signal; δ_{CT} is the "static" error caused by dead zone of the loop-controlled drive which, in turn, can be presented as:

$$\delta_{CT} = \delta_c + \delta_L + \delta_I, \tag{5.11}$$

where δ_c is the error component caused by friction forces and other resistance forces (friction forces, elastic forces and pressure forces of sealing bellows and membranes etc.); δ_L is the error component caused by recoil (gap in a kinematic chain) $\delta_D = \delta_3 + \delta_{11}$; δ_I is the instrumental error component determined by the accuracy rating of a sensor as well as by a zero drift of the amplifier.

The error of raster sensors used in the drives of electron beam lithography is comparable with the error tolerance (0.1 micron) typical for electron beam lithography. In the case of the use of a laser interferometer sensor, the instrumental error is two orders of magnitude smaller than "static" error and can be excluded from the total error. In the case of gapless transmission, which ensure $\delta_L = 0$, we can exclude this error component from the total error. Hydraulic drive has great advantages in this case because it usually has a very short kinematic chain (one shaft) and makes it possible to exclude dead zone ($\delta_L = 0$). The hydrodrive designed by the authors is supplied with the special sealing tube cuffs (see Figure5.2e) and can be used in oilless high vacuum. This hydrodrive has a high speed of response and high (for vacuum application) transporting speed, which can be varied in the range 10–100 mm/s. The special magnetic rheology drive (see Chapter 6) was designed by the authors to ensure the required combination of vacuum and mechanical properties. Considering the error components, it is necessary to note that component δ_N arises as a result of initial velocity V_N at the beginning of precise control mode, when the coordinate table reduces the high transporting speed into small speed for precise positioning.

So, the positioning error of a loop-controlled drive which works at stable outward parameters can be presented:

$$\delta\varphi_\Sigma = \delta_c + \delta_u. \tag{5.12}$$

It is known that the error component δ_c caused by resistance forces and called "dead zone" is determined by the "pickup current" (i), which depends on static resistant forces. So, the dead zone of a loop-controlled system is determined by general gain (carry-over factor) of measuring system:

$$\delta_c = \frac{i}{K_\partial \cdot K_y} = \frac{F_{mp}}{K_0}, \tag{5.13}$$

where K_δ is the gain of the sensor circuit; K_y is the gain of amplifier; F_{mp} is the static friction force, N; K_0 is the general gain of measuring system.

Hence, for the properly designed precise loop-controlled drive, the error of transference can be determined by the equation:

$$\delta\varphi_{\Sigma} = \frac{F_{mp}}{K_0} + \delta_u. \tag{5.14}$$

It is clear from this equation that even in the case of ideally precise measuring instruments, the error of transference will depend on the static resistance force (friction force) in the carrier of the drive. In traditional mechanical engineering the aim is to reduce the friction force F_{tp} using ball bearings, aerostatic or hydrostatic carriers. These designs cannot be used for vacuum technology because of, from the vacuum technology point of view, enormous gas flows from the carrier. The authors have designed the drive and the carriers which ensure very small resistance forces, which are less than the forces of the carriers mentioned above. By using matrix analysis, the new designs of precise magnetic rheology drives of nanometer precision were realized.

As it follows from Equation (5.14), for the error decreasing it is necessary to increase the general gain of measuring system K_0, but it leads to loss of speed of response of the system. It means that the value of K_0 must be chosen from certain optimum range.

5.3 Determination of the Error Components of Different Origins

5.3.1 Calculation of the Kinematic Component of the Error

Calculation of the kinematic component of the error of positioning is based on the constructional and technological analysis of the drive elements. It is based also on the analysis of kinematics of drive elements, on the analysis of technological error of drive elements and on the analysis of the errors resulting from drive elements mechanical interaction.

The method shown below gives a possibility to calculate the kinematic component of the positioning error of kinematic chains of vacuum mechanisms. The balance of components of resulting kinematic errors gives the possibility to see the dominating components and to analyze the possibility to decrease the total error. The method is demonstrated in this example:

1. The analytical model of the analyzed mechanism (drive) is created on the base of a kinematic scheme, general view and design drawings. The tolerances of form and sizes, the gaps, are shown in this model. These errors include the tolerances which determine the displacement of the components and units, and have influence on the forming resulting transference. These components and units are: ball bearings, shafts, gears, racks, guide screws, guides, etc., including different kinematic pairs are listed in the standards of kinematic pairs (for example, Russian Standards 1643-81, 1758-81, 9178-81, 9368-81, 9774-81, 10242-81, 13506-81, etc.). If the er-

ror depends on the size of a chain, the tolerances of the including components are shown and taken into account.

2. The frequencies (or periods) of all error components are determined. A detail (or unit) interaction with different components can create the errors with different frequencies and *vice versa*, different components situated in different parts of the mechanism can create errors with the same frequency of interaction, i.e. they have one frequency.

The errors with the same frequencies form one component of the total error of the mechanism. Mounting clearances at the mechanism work form the recoil which is calculated as a scalar sum. In the case, when the clearance value is the constant in different directions, the frequency of the corresponding error due to the clearance is equal to zero. In the case when the clearance value is different, for example in the case of eccentricity of gear ring or in the case of error of teeth pitch, the mounting clearance is a variable value.

The tolerances of the mounting components are considered as accidental values; the distribution law of every kind of the tolerance (error) is considered to be known. It gives a possibility to find the distribution parameters (average of distribution, dispersion, standard deviation) of every kind of the error. The components of every kind of the error are summarized as accidental vectors. This sum can be decreased using technological and constructional methods, for example, using selective assembling, joint processing or increasing the rigidity of the drive elements, etc.

3. The analytical dependencies, which interconnect the total transference error with every kind of elastic deformations and with every kind of technological errors, are developed.

4. The separate errors are summarized during the work cycles of the mechanism. Then, a graph of the total error as a function of the travel of an output element of the drive is made. The number of the cycles during the calculation sufficient for the reliable estimation of a kinematic error depends on the correlation of error frequencies and can be found by statistical methods. The total error calculated in such a way presents the "most technologically probable" variant of the mechanism. The statistical estimates of this error for "probable mechanism" or "probable drive" can be calculated also, for example: the largest for given probability error of the positioning of the drive, the probability of the error of given value, distribution function of the error, etc. The statistical estimations of the error for the lot of mechanisms or drives can be done using the same method by the separate errors summarizing and use of the variation coefficient which takes into account a dispersion of the element sizes in a mechanisms lot.

5. The partial influence of every component inside every harmony (frequency) is determined on the base of calculation results obtained according to points 2 and 3.

6. The recommendations how to decrease the kinematic error are developed through the analysis of the data obtained according to points 4 and 5.

This method is simple enough, but it does not give the possibility to determine the number of the characteristics of parameter reliability of the drive: the probability of no excess of given tolerance of the error for one mechanism, the probability of no excess of given tolerance of the error for the lot of mechanisms, etc.

The method described below of computer simulation does not have these problems. This method simulates the process of error forming for a separate unit (drive) as well as for a lot of units, which consist of an unlimited given number of units. This method simulates all known components of error of transference.

The method of computer simulation uses analytical model of the analyzed mechanism (drive) like analytical method described above. The form tolerances, the size tolerances, the gaps, which influence the resulting transference, are used in this model also. The frequency of occurrence of all error components, which were taken into account, are determined. The table with the tolerances, with the gaps, with the frequencies of their manifestation in corresponding error components as well as with the laws of distribution of the sizes within the tolerances is done.

The following functions are used for calculation and for the numerical presentation of the error of the components size and for determination of the gaps:

(a) in the case of a misalignment, the K parameter of Relay distribution is calculated:

$$K = 0.365\sqrt{\frac{2}{\pi}} \cdot D_r, \tag{5.15}$$

where D_r is the tolerance band for the misalignment of considered surfaces, microns;

(b) in the case of a gap the average of Gaussian distribution M and dispersion σ of distribution are calculated:

$$M(\delta\varphi_0) = \frac{(D^B - D^H)}{2} - \frac{(D^B_B - D^B_H)}{2}, \tag{5.16}$$

$$\sigma(\delta\varphi_0) = \frac{1}{6}[(D^B_0 - D^H_0)^2 + (D^B_B - D^H_B)^2]^{1/2}, \tag{5.17}$$

where D^B_0, D^B_B is the upper value of the hole, shaft tolerance, respectively in microns; D^H_0, D^H_B is the lower value of the hole, shaft tolerance, in microns.

For the simulation of the real error of transference in the drive it is necessary to simulate the process of real unit (drive) assembly. For the simulation of the real unit assembly it is necessary to use the array of random numbers which model the realization of real component errors, forms of detail, real gaps in the assembled unit as a result of the components manufacturing and of assembling of the unit. Repeating the process of unit assembly, it is possible to simulate the error modeling for the lot of the drives.

For the array of obtained random numbers it is possible to use the array of pseudo-accidental numbers, which can be generated using a special multiplication algorithm. Every next number X_{n+1} is generated from the previous one X_n using the equation:

$$X_{n+1} = (aX_n) \bmod C, \tag{5.18}$$

where a, c are the multiplier and the base of the number sequence, respectively; $(aX_n) \bmod C$ is the remainder of a X_n partition by C.

At the values $a = 3125$ and $C = 67108864$ and at the calculation precision of e^{-10} the number sequence is not converging. For the uniformly distributed number Y in the interval $[0 \ldots 1]$, it is possible to use the equation:

$$Y =_{n+1} /C. \tag{5.19}$$

The Gaussian-distributed number X_H may be obtained by the summation of $6 \ldots 8$ number Y.

For the given tolerance, the Gaussian distributed value of the size error can be calculated according to

$$X_H = \left(\sum_{i=1}^{6\ldots8} Y_i - 3 \right) \sqrt{2}\sigma + M. \tag{5.20}$$

The Relay distributed pseudo-accidental number X_p with K parameter can be obtained by the calculation the function Y:

$$X_p = K\sqrt{-\ln Y}. \tag{5.21}$$

It is necessary to note that the generated accidental number can be found exceeding the tolerance. (In reality, this component should be rejected by the test authority.) In the simulation method the accidental number generation is repeated to ensure the accidental size within the tolerance. This operation is advantageously different from the probable analytical method, which is based, in this case, on the use of truncated distribution laws.

The total amplitude A_j at frequency j can be calculated using the analytical model of the analyzed mechanism (drive) according to

$$A_j = \sqrt{A_{jk}^2 + A_{jy}^2}, \tag{5.22}$$

where A_{jx}, A_{jy} are the vector projections of the amplitude of the j-th harmony on the coordinate axes:

$$A_{jx} = \sum_{i=1}^{p} \xi_{ijx} r_{ij} \cos \alpha_{ij}, \quad A_{jy} = \sum_{i=1}^{p} \xi_{ijy} r_{ij} \sin \alpha_{ij}, \tag{5.23}$$

where r_{ij} is the generated accidental number, which simulates the size error of i-th component in the j-th harmony, microns; i is the error number in the j-th harmony; ξ_{ijx}, ξ_{ijy} are the influence (resp. the reduction) coefficients, which take into account the error reduction (multiplication) on the X, Y axes on the output element of the kinematic chain; α_{ij} is the angle of the i-th error vector which is uniformly distributed in the interval $[0 \ldots 360]$ angular degree.

This angle is also reduced to ξ_{ij} times and its value is simulated by

$$\alpha_{ij} = Y_i \cdot 360°/\xi'_{ij}. \tag{5.24}$$

The sum of current harmony errors (noted as $\delta\varphi_1$) can be determined for the mechanism in a general view:

$$\delta\varphi_1 = \sum_{j=1}^{n} A_j \cdot k_M \cdot (\sin(n_j\varphi + \varphi_{Hj}))/\xi'_{ij}, \tag{5.25}$$

where $\delta\varphi_1$ is the total harmony kinematic error simulated by computer; k_M is the coefficient of multiple contact, which takes into account the error decreasing because of transmitting features; ϕ current coordinate (the current angle of leading shaft); n_j is the multiplicity of j-th harmony appearing; φ_{Hj} is the initial phase of the vector A_j:

$$\varphi_{Hj} = \arctan(A_{yi}/A_{xi}). \tag{5.26}$$

In most cases, the coefficients of amplitude reduction $\varphi_{ijx(y)}$ on the X, Y axes are equal one to another. Then, the $\delta\varphi_1$ error may be found:

$$\delta\varphi_1 = \sum_{j}^{n}\sum_{i}^{p} r_{ij}\xi_{ij}k_M \sin((n_i\varphi + \varphi_{Hi})/\xi'_{ij}), \tag{5.27}$$

where $\xi_{ijx} = \xi_{ijy} = \xi_{ij}$.

The final result of component summation does not depend on the summation order. The final equation can be written as:

$$\delta\varphi_1 = \sum_{i=1}^{n} r_i k_M \xi_i \sin((n_i\varphi + \varphi_{Hi})/\xi'_{ij}), \tag{5.28}$$

where n is the total number of component of total error; $\xi_{ij} = 1/u_i$ is the transmitting ratio for the considered zone of the kinematic chain.

The recoil of the considered kinematic chain can be presented as an algebraic difference of real kinematic chain positions at the direct and at the backward travels of the mechanism. The recoil of the kinematic chain of the mechanism is also determined by the gaps, which depend on the form tolerances, the size tolerances, and the calculation principle is similar to the one, presented above but with different components. The recoil calculation differs from kinematic error calculation by the presence of only positive values of the error because it guarantees absence of the mechanism (drive) jamming. The recoil for the most coupling variants is a constant value, therefore the simulation algorithm uses two kinds of components: harmony component and constant component.

Let us consider the examples of kinematic error calculation for some examples of motion feedthrough, which are the most crucial elements of precise vacuum drives. Moreover, motion feedthroughs are the elements of the kinematic chain, which bring the largest part of the kinematic error because of its critical position on the border between atmosphere and vacuum.

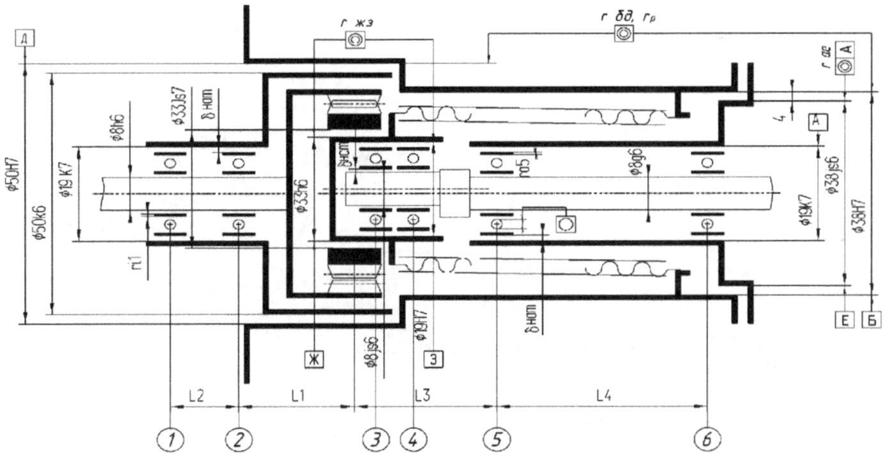

Fig. 5.5 The scheme of planetary gears feedthrough for the kinematic error calculation.

When the mechanism is simple, for example, ball bearing feedthrough, and has one coupling zone, its kinematic error directly influences the total error of transference of the drive.

Let us consider the calculation of the precision of the drive on four examples of vacuum mechanisms of ultrahigh vacuum equipment. According to position 1 of the described method, the analytical model of this four analyzed mechanisms was created:

- gear planetary rotary-motion feedthrough, Figure 5.5;
- bellows-sealed rotary-motion feedthrough, Figure 5.6;
- non-coaxial nut-screw linear motion feedthrough, Figure 5.7;
- linear screw drive with linearly deformed bellows, Figure 5.8.

The frequencies of different errors caused by the ball bearing inclination, are presented in Figure 5.2. The frequencies of different errors caused by the inclination of an eccentric, gears, nut-screw pair are presented in Figure 5.3. The errors with a similar nature are combined into different groups.

All periods of kinematic error harmonics are expressed through the period of the basic harmony. It is accepted that a basic period is equal to 360° turn of the leading shaft:

$$T_j = n \cdot 360^\circ / n_j, \tag{5.29}$$

where n is the frequency of the basic harmony (turn of leading shaft); n_j is the frequency of j-th harmony; T_j is the period of j-th harmony.

In all mentioned examples the amplitudes are considered as accidental values distributed with known distribution laws. Radial dimension drifts are recalculated in eccentricities.

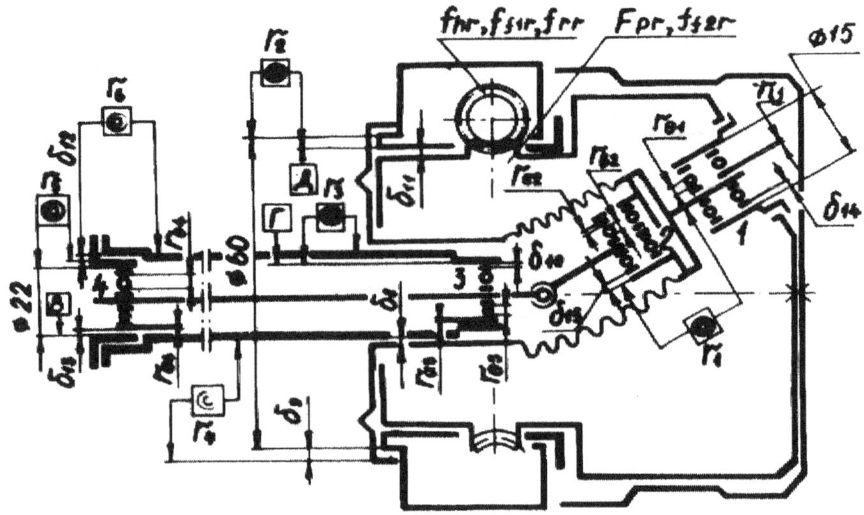

Fig. 5.6 The calculation scheme of angle shaft bellow feedthrough which is used for kinematic error calculation.

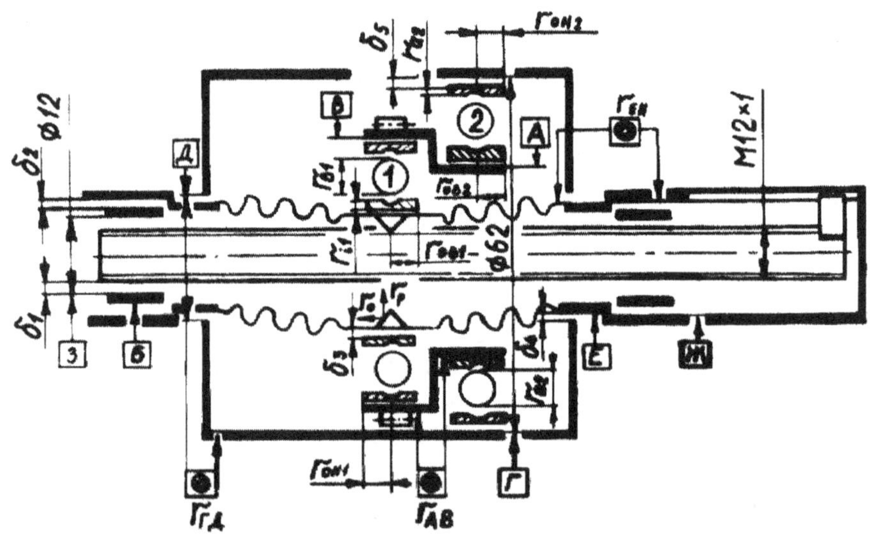

Fig. 5.7 The scheme of planetary nut-screw feedthrough for kinematic error calculation.

The average of distribution and dispersion of the error of every harmony are presented. The amplitude of every harmony is presented as a length of the vector, which is considered as a sum of accidental vectors of the considered errors.

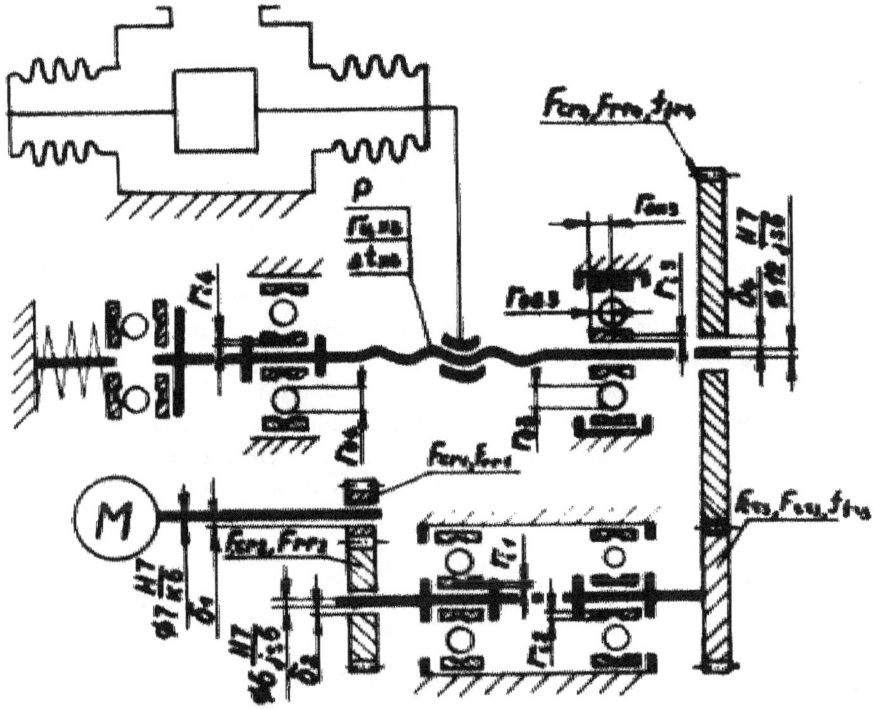

Fig. 5.8 The scheme of nut-screw drive for kinematic error calculation.

At summation operation all technological errors of considered harmony were corrected to the output element of the mechanism using linear transmitting ratios. At summation operation of independent accidental vectors distributed with different laws of distribution, the resulting vector parameters can be calculated according to the limit theorem. Average of distribution:

$$\begin{aligned} M(R_j) &= \pi/2\sqrt{\sum_{i=1}^{p} M(r_i^2)} = 0.886\sqrt{\sum_{i=1}^{p} M(r_i^2)}, \\ M(r_j^2) &= 1.27M^2(r_i), \end{aligned} \tag{5.30}$$

where $M(r_j^2)$ is the second moment.

Dispersion of the total vector value:

$$\sigma(R_j) = \sqrt{(1-\pi/4)\sum_{i=1}^{p} M(r_i^2)}. \tag{5.31}$$

At summation operation of independent accidental vectors, the vector of the sum is also the Relay accidental vector with average distribution:

$$M(R_j) = \sqrt{\sum_{i=1}^{p} M^2(r_i)}. \quad (5.32)$$

The calculated amplitudes of errors of gear planetary rotary-motion feedthrough and bellows-sealed rotary-motion feedthrough were recalculated into angular errors of output shafts. The calculated amplitudes of non-coaxial nut-screw linear-motion feedthrough and linear screw drive with linearly deformed bellows were recalculated into linear errors of output components.

The average of distribution and dispersion of the error of every harmony were calculated for all mentioned mechanisms; the amplitudes of the different errors were summed as independent accidental values because they are aliquant and differ noticeable one from another. The probable estimations of their output parameters were calculated:

$$M(\delta\varphi) = 2\sum_{j=1}^{n} M(R_j), \quad (5.33)$$

$$\sigma(\delta\varphi) = 2\sqrt{\sum_{j=1}^{n} \sigma^2(R_j)}, \quad (5.34)$$

where $M(\delta\varphi)$, $\sigma(\delta\varphi)$ are the averages of distribution and dispersion of the kinematic error of positioning of the output element of the mechanism, respectively; R_j is the considered initial error.

The averages of distribution and dispersion of the kinematic error of positioning of the output element for the four considered mechanisms are:

- gear planetary rotary-motion feedthrough: $M(\delta\varphi_1 = M(F_{\max}) = 15.8$ ad, $\sigma(\delta\varphi_1) = \sigma(F_{\max}) = 5.6$ ad;
- non-coaxial nut-screw linear-motion feedthrough: $M(\delta\varphi_1 = M(F_{\max}) = 28.22$ microns, $\sigma(\delta\varphi_1) = \sigma(F_{\max}) = 12.22$ microns;
- bellows-sealed rotary-motion feedthrough: $M(\delta\varphi_1 = M(F_{\max}) = 28.22$ ad, $\sigma(\delta\varphi_1) = \sigma(F_{\max}) = 11.83$ ad;
- linear screw drive with linearly deformed bellow: $M(\delta\varphi_1 = M(F_{\max}) = 17.7$ microns, $\sigma(\delta\varphi_1) = \sigma(F_{\max}) = 5.56$ microns,

where $F_{\max}$ is the maximum kinematic error.

The recoil error, denoted by symbol δ_3, for the mentioned mechanisms (gear planetary rotary-motion feedthrough, bellows-sealed rotary-motion feedthrough, non-coaxial nut-screw linear motion feedthrough, linear screw drive with linearly deformed bellow) was calculated as a sum of independent accidental gaps adjusted to the output element.

The average of distribution and dispersion of the kinematic error of recoil of the output element of the mechanisms was calculated:

$$M(j_{\varphi\Sigma}) = \sum_{i=1}^{n} M(j_{\varphi_i}),$$

$$\sigma(j_{\varphi\Sigma}) = \sqrt{\sum_{i=1}^{n} \sigma^2(j_{\varphi_i})}, \tag{5.35}$$

where $j_{\varphi\Sigma}$ is the i-th gap error adjusted to the output element of the drive.

The recoil error for the gear planetary rotary-motion feedthrough and bellows-sealed rotary-motion feedthrough were recalculated into angular errors of output details.

The figures of recoil error for the mentioned mechanisms are:

- non-coaxial nut-screw linear motion feedthrough: $M(\delta_3) = 32.6$ microns, $\sigma(\delta_3) = 2.85$ microns;
- gear planetary rotary-motion feedthrough: $M(\delta_3) = 24$ ad, $\sigma(\delta_3) = 2.68$ ad;
- bellows-sealed rotary-motion feedthrough: $M(\delta_3) = 1.35$ ad, $\sigma(\delta_3) = 11.33$ ad.

Balance of error component of non-coaxial nut-screw linear motion feedthrough shows that the largest influence on the resulting kinematic error has the sizes of the ball bearing (radial runout $r_{OH1}, r_{OB1}, r_{OH2}, r_{OB2}$) and the gap δ_4 (see designation in Figures 5.7–5.8). The partial influence (in percentage) of different harmonies on the total kinematic error of every mechanism is shown in Tables 5.2–5.6.

The way to control kinematic error using the error balance of the mechanism can be illustrated by the following example.

Figure 5.9a shows the harmony variation (with amplitudes $A_1 \dots A_i$) of different errors of ultrahigh vacuum harmonic rotary feedthrough (see Figure 2.2) as a function of leading shaft turning φ. Figure 5.9b shows the balance of these harmonies. It can be seen that the first harmony A_1 (see Figure 5.9a) is dominating and it is determined by the technological precision of the sizes 38H7/js6 processing which determine the value of this gap.

The probable approach to the parametrical reliability estimation gives a possibility to create the balances of error component distribution on the stage of design. Also, it gives a possibility to estimate the parametrical reliability (η_n coefficient) depending of tolerance of resulting error value.

Figure 5.10 shows the probability $P(|\delta\varphi| < \Delta L)$ of the tolerance ΔL exceeded by the total kinematic error for the different qualities of the ball bearing precision of non-coaxial nut-screw linear motion feedthrough (see Figure 2.5). Figure 5.10 shows that when the tolerance is $\Delta L = 20$ microns, it is necessary to use the ball bearing of the second quality; when the tolerance is $\Delta L = 50$ microns it is necessary to use the ball bearing of the fifth quality.

Experimental research of precision of non-coaxial nut-screw linear-motion feedthrough shows that the output load increasing decreases the error cycling variation (Figure 5.11), decreases some elastic error components and increases the backlash component.

Table 5.2 The frequencies of kinematic errors of different vacuum mechanisms: Planetary gear feedthrough,bellow feedthrough, planetary nut-screw feedthrough, nut-screw with axial bellow deformation.

Source of error		Error		No. of ball bearings	Planetary gear feedthrough (A)	Bellow feedthrough (B)	Planetary nut-screw feedthrough (C)	Nut-screw (D)
Ball bearings	Rings	Ball traces inclination	Static ring	1	n	n	$n\frac{d_{o1}-D_{T1}Cos\alpha_1^*}{2d_{01}};n$	$n\frac{z_1}{z_2}\frac{d_{o1}-D_{T3}Cos\alpha_1^*}{2d_{01}}$
				2	n	n	$n\frac{d_{o2}-D_{T2}Cos\alpha_2^*}{2d_{02}};n$	$n\frac{z_1}{z_2}\frac{d_{o2}-D_{T2}Cos\alpha_2^*}{2d_{02}}$
				3, 4	n	n	—	$n\frac{z_1}{z_2}\frac{z_3}{z_4}\frac{d_{o3}-D_{T3}Cos\alpha_3^*}{2d_{03}}$
			Moving ring	1	$n\frac{z_1}{z_2}$	—	$n\frac{d_{o1}-D_{T1}Cos\alpha_1^*}{2d_{01}}$	$n\frac{z_1}{z_2}\frac{d_{o1}-D_{T1}Cos\alpha_1^*}{2d_{01}};n\frac{z_1}{z_2}$
				2	—	—	$n\frac{d_{o2}-D_{T2}Cos\alpha_2^*}{2d_{02}}$	$n\frac{z_1}{z_2}\frac{d_{o2}-D_{T2}Cos\alpha_2^*}{2d_{02}};n\frac{z_1}{z_2}$
				3, 4	—	—	—	$n\frac{z_1}{z_2}\frac{z_3}{z_4}\frac{d_{o1}-D_{T1}Cos\alpha_1^*}{2d_{01}}$
	Balls	Size deviation		1	$n(1-\frac{z_1-z_2}{z_2}\frac{(d_{o1}-D_{T1})}{2d_{01}})z_{u1}$	$n\frac{(d_{o1}-D_{T1})}{2d_{01}}z_{u1}$	$n\frac{(d_{o1}-D_{T1}Cos\alpha_1)}{2d_{01}}z_{u1}$	$n\frac{z_1}{z_2}\frac{(d_{o1}-D_{T1}Cos\alpha_1)}{2d_{01}}z_{u1}$
				2	$n\frac{(d_{o2}+D_{T2})}{2d_{02}}z_{u2}$	$n\frac{(d_{o2}+D_{T2})}{2d_{02}}z_{u2}$	$n\frac{(d_{o2}-D_{T2}Cos\alpha_2)}{2d_{02}}z_{u2}$	$n\frac{z_1}{z_2}\frac{(d_{o2}-D_{T2}Cos\alpha_2)}{2d_{02}}z_{u2}$
				3, 4	$n\frac{(d_{o3}+D_{T3})}{2d_{03}}z_{u3}$	$n\frac{(d_{o3}+D_{T3})}{2d_{03}}z_{u3}$	—	$n\frac{z_1}{z_2}\frac{z_3}{z_4}\frac{d_{o3}-D_{T3}Cos\alpha_3}{2d_{03}}z_{u3}$
		Defect of roundness		1	$n\frac{z_1-z_2}{z_2}\frac{(d_{o1}^2-D_{T1}^2)}{d_{01}D_{T1}}$	$n\frac{(d_{o1}^2-D_{T1}^2)}{d_{01}D_{T1}}$	$n\frac{(d_{o1}^2-D_{T1}^2Cos^2\alpha_1)}{d_{01}D_{T1}}$	$n\frac{z_1}{z_2}\frac{(d_{o1}^2-D_{T1}^2Cos^2\alpha_1)}{d_{01}D_{T1}}$
				2	$n\frac{(d_{o2}^2-D_{T2}^2)}{d_{02}D_{T2}}$	$n\frac{(d_{o2}^2-D_{T2}^2)}{d_{02}D_{T2}}$	$n\frac{(d_{o2}^2-D_{T2}^2Cos^2\alpha_2)}{d_{02}D_{T2}}$	$n\frac{z_1}{z_2}\frac{(d_{o2}^2-D_{T2}^2Cos^2\alpha_2)}{d_{02}D_{T2}}$
				3, 4	$n\frac{(d_{o3}^2-D_{T3}^2)}{d_{03}D_{T3}}$	$n\frac{(d_{o3}^2-D_{T3}^2)}{d_{03}D_{T3}}$	—	$n\frac{z_1}{z_2}\frac{z_3}{z_4}\frac{(d_{o3}^2-D_{T3}^2Cos^2\alpha_3)}{d_{03}D_{T3}}$

* Axial error

Table 5.2 Continued.

ɔurce	Error	(A)	Error	(B)	Error	(C)	Error	(D)
:ntricity	$r_{БД}$, r_{AE}	n	r_1, r_2, r_3	n	$r_{EЖ}$, $r_{ГД}$	n	—	—
	r_{34}, $r_{Ж3}$	n	r_5, r_6, r_3	n	$r_{3Б}$	n	—	—
	$d_1, \neg 60$	n	d_8, d_9	n	$d_4, \neg 25$	n	$d_1, \neg 7$	n
	$d_2; \neg 24,2$	n	d_{11}, d_{12}	n	—	—	$d_2, \neg 6$	nz_1/z_2
	—	—	d_{13}	n	—	—	$d_4, \neg 12$	nz_1z_3/z_2z_4
nclination,	F_{rr1}, F_{cr1}	n	f_{rr}, f_{hr}	n	—	—	F_{rr1}, F_{cr1}	n
	F_{rr2}, F_{cr2}	n	—	—	—	—	F_{rr2}, F_{cr2}	nz_1/z_2
:ation	—	—	—	—	—	—	F_{rr3}, F_{cr3}	nz_1/z_2
rror	—	—	—	—	—	—	F_{rr4}, F_{cr4}	nz_1z_3/z_2z_4
ror of	f_{fr1}, f_{fr2}	nz_2	f_{f1r}, f_{f2r}	n	—	—	f_{fr1}, f_{fr2}	nz_1
s shape	—	—	—	—	—	—	f_{fr3}, f_{fr4}	nz_1z_3/z_2
cal error	—	—	—	—	—	—	—	nz_1z_3/z_2z_4
radial	—	—	—	—	r_{p2}	n	—	—
axial	—	—	—	—	r_{o2}	n	—	—

Table 5.3 The errors considered in a process of hermetic planetary feedthrough calculation (for output shaft). r_p – axis inclination of surfaces B and D of the body. Axis inclination after body manufacturing $r_{b\partial}$.

	Error (x)	Distribution parameters				Distribution low	n_i
		M(x)	D(x)	σ(x)	$M(x^2)$		
1	$r_{бд}$	3,29	2,95	1,71	13,8	Relay	n1
2	r_{ae}	3,29	2,95	1,71	13,8		n1
3	$r_{5,6}$	1,76	0,85	0,92	3,92		n1
4	$r_{жз}$	2,2	1,32	1,15	6,14		n1
5	r_{a1}	1,61	0,71	0,84	3,28		n1
6	r_{a2}	3,07	2,57	1,60	12,00		n1
7	r_{a3}	1,46	0,58	0,76	2,74		n1
8	r_{a4}	1,46	0,58	0,76	2,74		n1
9	r_{a5}	2,47	1,67	1,29	7,76		n1
10	r_{a6}	1,01	0,28	0,53	1,61		n1
11	Fr_2	2,92	2,33	1,53	10,86		n1
12	F_{c2}	1,46	0,58	0,76	0,67		n1
13	r_p	140	—	—	—	—	n1
14	δ∅50H7/k6	2,50	23,96	4,89	30,21	Normal	n1
15	δ∅38H7/js6	12,50	23,96	4,89	180,21		n1
16	δ∅33Js7/h6	8	6,96	2,64	70,96		n1
17	δ∅19H7	14,5	14,72	3,84	—		—
18	δ_1∅8h6	1,1	5,32	2,31	—		—
19	δ_2∅8h6	2,1	19,40	4,40	—		—
20	δ_5∅8g6	10,14	12,57	3,55	—		—
21	δ_6∅8g6	4,14	2,09	1,45	—		—
22	$\delta п_1$	11,00	21,78	4,67	—		—
23	$\delta п_2$	21,00	79,38	8,91	—		—
24	$\delta п_3$	4	7,7	2,77	—		—
25	$\delta п_4$	4	7,7	2,77	—		—
26	$\delta п_5$	16,90	51,41	7,17	—		—
27	$\delta п_6$	6,90	8,57	2,93	—		—
28	$j_п$	21,00	15,67	3,96	—		—
29	ri_1	1,20	0,40	0,63	—	Relay	n2
30	ri_2	2,30	1,44	1,20	—		n2
31	Fr_1	2,92	2,33	1,53	—		n2
32	Fc_1	1,46	0,58	0,76	—		n2
33	Θ_1	2,01	1,10	1,05	—		n3
34	Θ_2	3,84	4,01	2,00	—		n3
35	Θ_3	1,83	0,91	0,95	—		n4
36	Θ_4	1,83	0,91	0,95	—		n4
37	Θ_5	3,09	2,60	1,61	—		n5
38	Θ_6	1,26	0,43	0,66	—		n5
39	f_{f1}	2,56	1,79	1,34	—		n6
40	f_{f2}	2,56	1,79	1,34	—		n6

Table 5.4 The errors considered at the calculation of angular shaft bellow feedthrough. The figures are shown for output shaft.

No.	Total error	Distribution law	Distribution law parameters			Period	Contribution
			$M(X_i)$	$S(X_i)$	$M(X_i^2)$		
1	δ'_8		11,5	6,2	170,68		
2	δ'_9		60	20	4000		
3	δ'_{11}	Normal	25	8,3	693,9		
4	δ'_{12}		18,5	6,2	380,7		
5	δ'_{13}		16	5,33	284,4		
6	r_1		1,83	0,95	4,22		
7	r_2		3,65	1,9	16,94		
8	r_3		3,65	1,9	16,94	T_1=360	77,1%
9	r_4		3,65	1,9	16,94		
10	r_5		1,82	0,95	4,22		
11	r_6		3,65	1,9	16,94		
12	r_{i1}		1,47	0,77	2,74		
13	r_{A2}		1,47	0,77	2,74		
14	r_{A3}		1,47	0,77	2,74		
15	r_{A4}		21	7	490		
16	F_{pr}	Relay	1,82	0,95	4,22		
17	f_{f2r}		3,63	1,89	-		
18	f_{hr}		4,36	2,27	-	T_2=6	1,6%
19	f_{f1r}		5,5	2,9	-		
20	f_{rr}		2,1	1,1	-		
21	$r_{i5}=r_{i6}$		1,83	0,96	-		
22	$r_{\theta 5,6}$		1,83	0,96	-	T_3=3	0,2%
23	$r_{\theta 1}$		1,83	0,96	-	T_4=172	13,3%
24	$r_{\theta 2}$		1,83	0,96	-	T_5=254	6,1%
25	$r_{\theta 3,4}$		1,83	0,96	-	T_6=205	1,7%
26	δ'_{10}		2,5	0,83	-		
27	δ'_{14}		13,5	4,45	-		
28	δ'_{15}		16	5,3	-		
29	δ_5		13,5	4,45	-		
30	$\delta_{П1}$	Normal	10,5	1,83	-	Люфт	
31	$\delta_{П2}$		10,5	1,83	-		
32	$\delta_{П3}$		10,5	1,83	-		
33	$\delta_{П4}$		10,5	1,83	-		
34	$\delta_{П5}=\delta_{П6}$		10,5	1,83	-		
35	$\delta_{3ц}$		138	6	-		

The mechanisms with one zone of coupling have large specific contacting loads that lead to high wear rate in the contacting zone in vacuum. Irregular wear increases the kinematic component of the total error and its dispersion as shown in Figure 5.12. The experimentally obtained equation shows the average error of positioning of non-coaxial nut-screw linear motion feedthrough as a function of cycles of load number N:

Table 5.5 The errors which can be detected at calculation of planetary nut-screw feedthrough. The calculated results are collected to outgoing shaft.

No.	Error title.	Distrib-ution law	Distribution Law Parameters			Period, degrees	Contrib-ution
			$M(X_i)$	$S(X_i)$	$M(X_i^2)$		
1	δ'_4	Normal	8,5	22,04	76,4	T_1=360	43,8%
2	$r_{\text{ГД}}$	Relay	3,0	1,06	11,46		
3	$r_{\text{ЖЕ}}$		2,0	1,04	76,41		
4	$r_{\text{ЗБ}}$		2,0	1,04	11,46		
5	r_{i1}		1,8	0,94	509		
6	r_{A2}		3,65	1,91	509		
7	r_{p2}		3,65	1,91	4,13		
8	r_{O2}		1,8	0,94	16,96		
9	r_{OB1}		1,46	2,09	16,96	T_2=634	10,3%
10	r_{OB2}		1,46	2,09	4,13	T_3=822	10,3%
11	r_{OH1}		2,01	2,88	-	T_4=830	14,1%
12	r_{OH2}		2,01	2,88	-	T_5=640	14,1%
13	$r_{\theta 1}$		0,9	0,47	-	T_6=48,5	3,7%
14	$r_{\theta 2}$		0,9	0,47	-	T_7=45,5	3,7%
15	r_{AB}		3,0	2,5	-	Gap	
16	r_{AB}		1,0	0,1	-		
17	δ_1	Normal	18	2,08	-		
18	δ_2		17	3,97	-		
19	δ_3		16	6,04	-		
20	δ_5		21,5	2,33	-		
21	$\delta_{\text{П}1}$		16,4	2,37	-		
22	$\delta_{\text{П}2}$		14,8	2,73	-		

$$\delta L = \delta_3 + \delta_4 = 16 + 62.5 \times 10^{-5}\,\text{N, microns.} \tag{5.36}$$

The research of backlash of non-coaxial nut-screw linear-motion feedthrough shows, that the average of this error (denoted δ_3) linearly depends on the number of the cycles as shown in Figure 5.13.

The instability of the wear process in vacuum leads to increase of the dispersion of the positioning error. This error component denoted as δ_4 is clearly visible in Figure 5.13. This error component is larger than the δ_3 component.

Let us consider another components of the total error on the example of non-coaxial nut-screw linear motion feedthrough. Under a ten-fold axial load increase its total error decreases from 99 to 44 microns (see Figure 5.11); from the other side, after 100000 working cycles its total error increases from 20 to 50 microns. At that time, the backlash increases from 25 to 70 microns. Remember for comparison, that the kinematic error of non-coaxial nut-screw linear motion feedthrough caused by technological reasons (sizes errors, gaps, forms errors) is 30–60 microns.

Thus, we can see that the designer can calculate the allowable tolerances of the elements of the designed mechanism to ensure its parametrical reliability on the precision parameter.

Table 5.6 The errors which are calculated for nut-screw feedthrough with axial bellow deformation. The error parameters are calculated for outgoing shaft.

No.	Error title	Distribution law	Parameters of distribution			Period angular degree	Contribution
			$M(X_i)$	$S(X_i)$	$M(X_i^2)$		
1	δ'_1	Normal	0,023	0,033	0,0016		
2	F_{rr1}	Relay	0,036	0,019	0,0016	T_1=59,4	1,4%
3	F_{cr1}		0,107	0,056	0,0146		
4	δ'_2	Normal	0,068	0,027	0,0053		
5	F_{rr2}		0,099	0,052	0,0125		
6	F_{cr3}		0,0223	0,117	0,0633		
7	F_{i1}	Relay	0,018	0,009	0,0004	T_2=123	4,3%
8	F_{cr3}		0,074	0,039	0,007		
9	r_{i1}		0,157	0,082	0,0314		
10	r_{i2}		0,201	0,105	0,0514		
11	δ'_4	Normal	0,211	0,085	0,0512		
12	F_{rr4}		0,233	0,177	0,632		
13	F_{cr4}		0,056	0,029	0,004	T_3=360	8,9%
14	r_{i3}		0,095	0,05	0,0114		
15	r_{i4}	Relay	0,009	0,005	0,0001		
16	$r_{цхв}$		0,73	0,382	0,6784		
17	f_{fr3}		0,111	0,058	-	T_4=8,8	1,6%
18	f_{fr4}		0,086	0,045	-		
19	$r_{\theta 3}$		0,026	0,0135	-	T_5=79,8	0,3%
20	$r_{\theta 4}$		0,001	0,0005	-		
21	r_{OH3}		3,7	1,94	-	T_6=961	41,8%
22	r_{OB3}		3,7	1,94	-	T_7=577	41,8%

From the other side, we can see that to ensure the parametrical reliability on the precision parameter the designer has to take into account not only kinematic components of the total error, but also other components which can be dominating.

5.3.2 Calculation of the Error from Elastic Deformations

The specific feature of vacuum equipment is the presence of the increased number of different forces, which act on the vacuum mechanism. The necessity to use ultrahigh vacuum equipment increases requirements to sealing elements. Stainless bellows, membranes, flexible steel elements can be used as sealing elements. The deformation of these elements causes of appearing elastic forces; the atmosphere pressure difference causes forces of atmosphere pressure; the vacuum variation causes friction coefficient and friction force variation. The necessity of high temperature baking for ultrahigh vacuum mechanisms leads to use at large gaps in the designed mechanisms to avoid its jamming. These gaps create additional component of total error of positioning. To eliminate these errors, the designers use different compen-

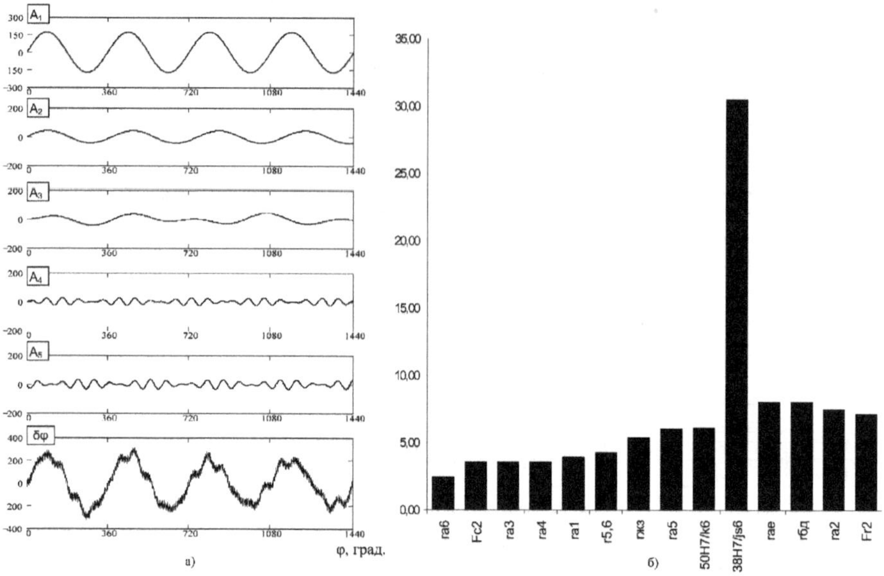

Fig. 5.9 The errors which are used for the precision of harmonic drive calculation. (a) The kinematic error forming $\delta\varphi$ of hermetic planetary drive. A_i $(i = 1 \ldots 6)$ are the components of summarized error based on frequency n_i (the component A_6 is not shown); φ is the rotation angle of the leading shaft. (b) The errors balance based on frequency n_1. The error titled 38H7/js6 is based on the base sizes.

sators, springs, devices for backlash elimination, which create additional forces and, in turn, create additional elastic error component.

So, we can see that balance of the errors of positioning of precise vacuum mechanisms is different from the traditional mechanical engineering and the described method gives a possibility to designer to control this balance and to ensure the required precision.

5.3.3 Calculation of the Error Caused by the Deformation of the Thin-Wall Sealing Elements

It is necessary to note that the thin-wall sealing element during the work in a motion feedthrough undergoes circular deformation that causes harmony error because of sealing element elasticity anisotropy and thin wall thickness instability. In a case of linear deformation of thin-wall sealing element it causes linearly varied error because of linearly varied elastic forces.

Let us consider this error component in rotary-motion feedthroughs. The size errors at manufacturing of thin-wall elements and its displacements at fixation point on a base cause elastic forces, which have harmonic character. So, the elastic error

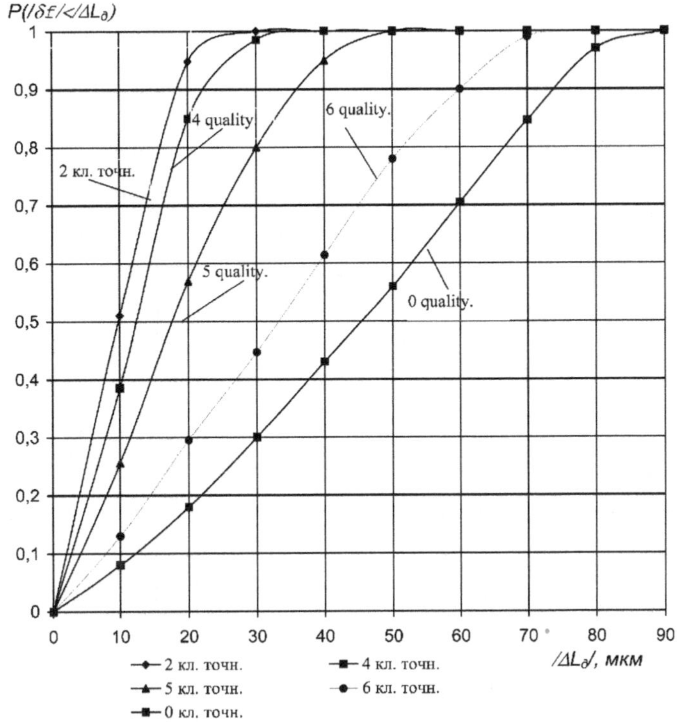

Fig. 5.10 The probability $P(\delta\varphi < \Delta L_\partial)$ of planetary nut-screw feedthrough that the error of feedthrough $(\delta\varphi P(\delta\varphi < \Delta L_\partial)$ is smaller than the error value of maximal considered error ΔL_∂ for different quality precision of ball bearings.

component δ_9 (see Table 5.1) due to of displacement fixation of the sealing bellows on the feedthrough body (see Figure 5.14), can be expressed as

$$\delta_9 = (J_u \cdot e \cdot \Delta_0 \cdot \xi_{i1}/J_{i1}) \sin((\varphi_e + \varphi_{ne})/\varphi_{ij}), \tag{5.37}$$

where J_u is the bend rigidity of the bellows; e is the work eccentricity (eccentricity of the shafts) of the feedthrough; Δ_0 is the bellows mounting eccentricity (displacement of mounting) on the body; J_{i1} is the torsion rigidity of the drive of the kinematic chain part from the beginning to i-th (to the feedthrough); φ_e, φ_{ne} are the turning and initial angles of a leading shaft of feedthrough, respectively.

The elastic error component δ_9 of harmonic rotary feedthrough due to displacement fixation of the sealing thin-wall tube (with flexible gear) on the feedthrough body (see Figure5.15) can be expressed as

$$\delta_9 = (J_u \Delta_0 \Delta_\Gamma \cdot 2\varepsilon_i/J_{i1}) \sin((\varphi_\Gamma + \varphi_{H\Gamma})/\varepsilon'_{i1}), \tag{5.38}$$

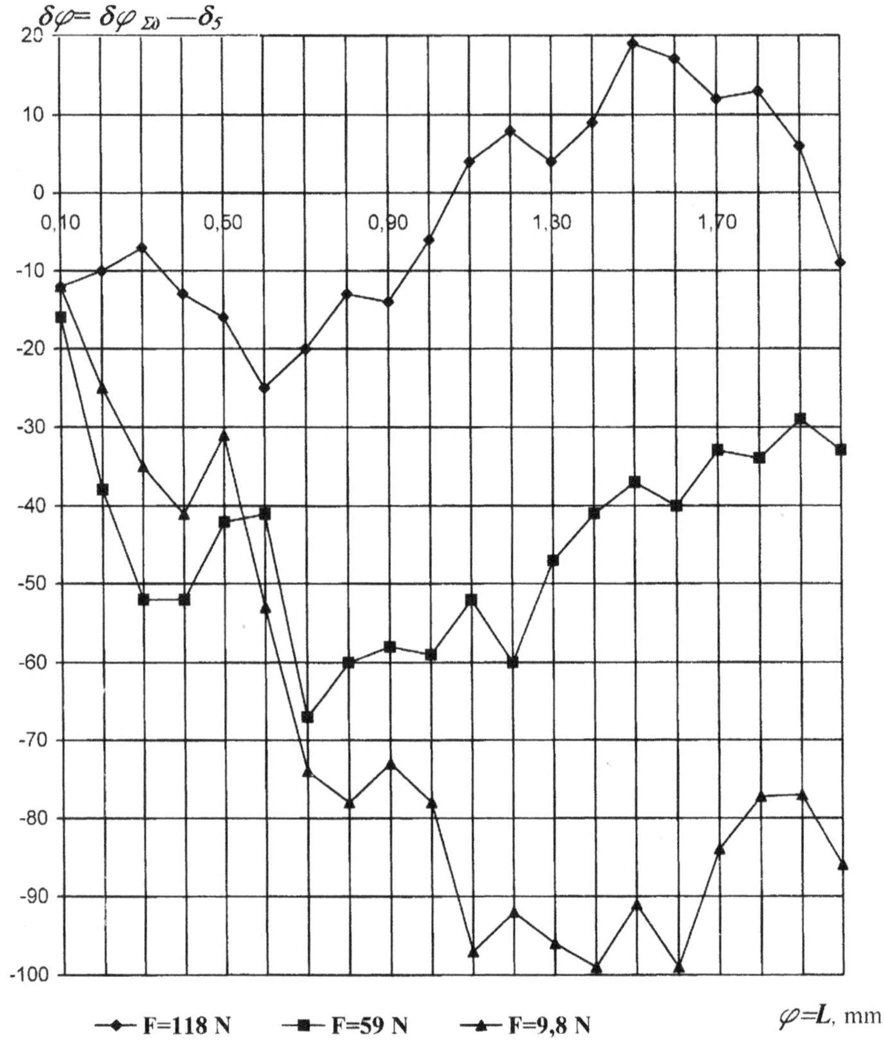

Fig. 5.11 Experimental research of precision of non-coaxial nut-screw linear-motion feedthrough $\delta\varphi$ under various axial loads as a function at linear transference. The elastic deformation (δ_5) is rested.

where φ_Γ, $\varphi_{H\Gamma}$ are the turning and initial angles of a leading shaft (deformation generator) of harmonic rotary feedthrough, respectively; Δ_0, Δ_Γ are the displacements of mounting of thin-wall tube (with flexible gear); J_I is the bend rigidity of the thin-wall tube; ε is the accidental number uniformly distributed in the range $[0 \ldots 1]$.

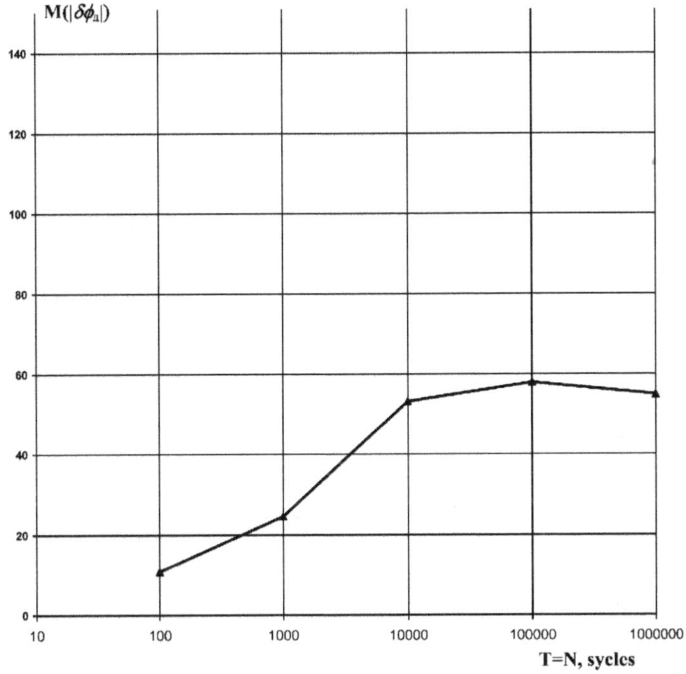

Fig. 5.12 The real positioning error variation $M(|\delta\varphi_D|)$ of planetary nut-screw feedthrough as a result of working cycles.

The elastic error component δ_9 of harmonic rotary feedthrough due to elasticity anisotropy of a thin-wall tube (with flexible gear) because of thin wall thickness instability, see Figure 5.15, may be expressed as

$$\delta_9 = (J_r^2(\Delta_h)^2\varepsilon_i/2 \cdot J_I J_i)\sin((u\varphi_\Gamma + \varphi_{H\Gamma})/\varepsilon'_{i1}), \qquad (5.39)$$

where u is the number of coupling zones ($u = 2$); j_r is the radial rigidity of the of thin-wall tube; Δh is the thin wall thickness instability.

Equations (5.37–5.39) show that the errors have a harmonic nature. These errors can be summed with the kinematic error components using a single mathematical operator according to Equation (5.28).

For the parametrical reliability (probability $P(|\delta\varphi| < \Delta L)$) estimation on the stage of design it is necessary to calculate also dispersions of mentioned errors using Equations (5.37, 5.38, 5.39).

During the linear travel of mechanical elements in vacuum the sealing elements (bellows, membranes) and the devices for gaps elimination (springs) create the additional forces, which linearly vary during the transference.

So, analysis of the elastic parameter of sealing welded bellows (Figure 5.16) shows that elastic component δ_9 of the error can be adopted as a linearly varied

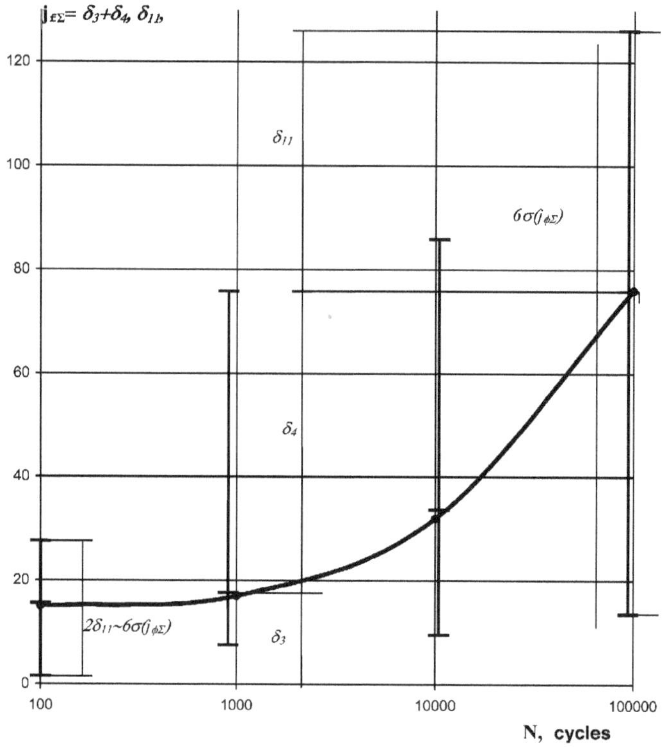

Fig. 5.13 The diagram of error components $\delta_3 + \delta_4$ variation (dead stroke) $\mathbf{j}_{f\Sigma}$ and of error in result of details mistakes δ_{11} for planetary nut-screw feedthrough as a result of its work in vacuum.

function in certain regions of transference zone:

$$\delta_9 = \delta\varphi_L = \sum_{k=1}^{c}\sum_{i=1}^{n} J_K \delta\varphi_k \varepsilon_i / J_{i1}, \tag{5.40}$$

where J_k is the rigidity of a sealing element (bellow) on the k-th region of its deformation; J_{i1} is the torsion rigidity of the drive of the kinematic chain part from the beginning to i-th; $\delta\varphi_k$ is the transference of the working part of the elastic element (bellow) on the k-th region of its deformation.

When we have non-linear variation of the resistance force, the elastic component δ_9 of the positioning error can be expressed as a function of energy E_k variation spent on the deformation of the elastic elements of kinematic chain:

$$\delta_9 = \sum_{k=1}^{c}\sum_{i=1}^{n} \frac{\partial E_k}{\partial(\varphi/\xi_u)} \frac{\varepsilon_i}{J_{i1}}. \tag{5.41}$$

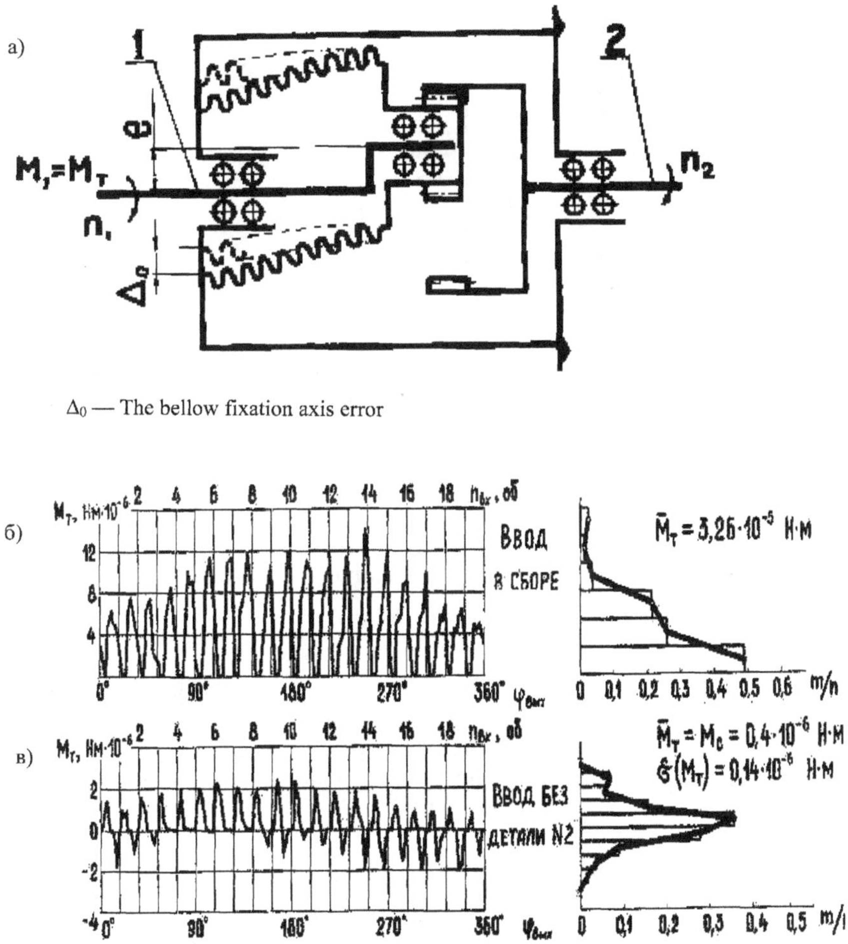

Fig. 5.14 The scheme of elastic error forming as a result of bellow axis inclination (D_0). In the planetary gears rotary-motion feedthrough: (a) scheme of the resistant moment forming: 1 – leading shaft; 2 – outlook shaft; (b) the diagram of resistant moment variation in the assembled feedthrough.

In special cases, the elastic component δ_9 of the positioning error can be expressed as a probable, harmonic, or linear function.

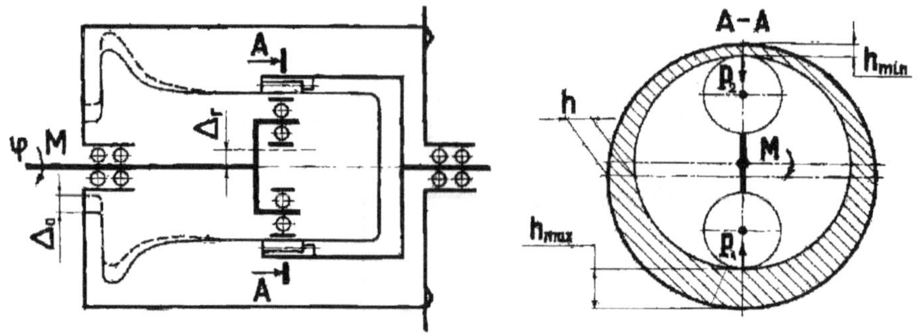

Fig. 5.15 The scheme of elastic error of positioning forming as a result of thin-wall element thickness difference. Δ_0, Δ_Γ – the mounting eccentricity of thin-wall element and generator, respectively. h – the base thickness of a thin-wall element. h_{min}, h_{max} – the minimal and maximal thickness of the thin-wall element. P_1 and P_2 – the forces which act to the generator discs. M – the leading shaft moment.

5.3.4 Calculation of the Positioning Error Caused by the Resistance Forces at Movement

In addition to elastic forces in a vacuum drive, vacuum intensifies the resistance forces of accidental nature appearing during motion. This can cause also additional elastic error components. Below we consider the mechanics of this forces (and errors) on the examples of vacuum ball bearing and non-coaxial nut-screw ultrahigh vacuum drive.

- *Error component due to the resistance torque of ball bearing fluctuation*

The phenomenon of the resistance torque of ball bearing fluctuation considered below takes place as in vacuum also as in traditional mechanical engineering. However, in traditional mechanical engineering it is negligible while it has critical importance in high and ultrahigh vacuum, where friction coefficient grows upto high level. Combination of high resistance torque of ball bearing fluctuation simultaneously with small rigidity of kinematic chain elements of vacuum mechanisms causes the appearance of the new error component, which is accidental in its form and is elastic in its nature.

The reason of this resistance torque is the presence of small difference in the sizes of balls (0.1–0.5 microns for different qualities). In case of axial load when all balls are squeezed between inner and outer rings, different balls have different orbital speed because of different in sizes. The "fast" balls ("leading" balls) push forward the retainer of the ball bearing, the "slow" balls ("driven" balls) are retarded by the retainer and act as brakes.

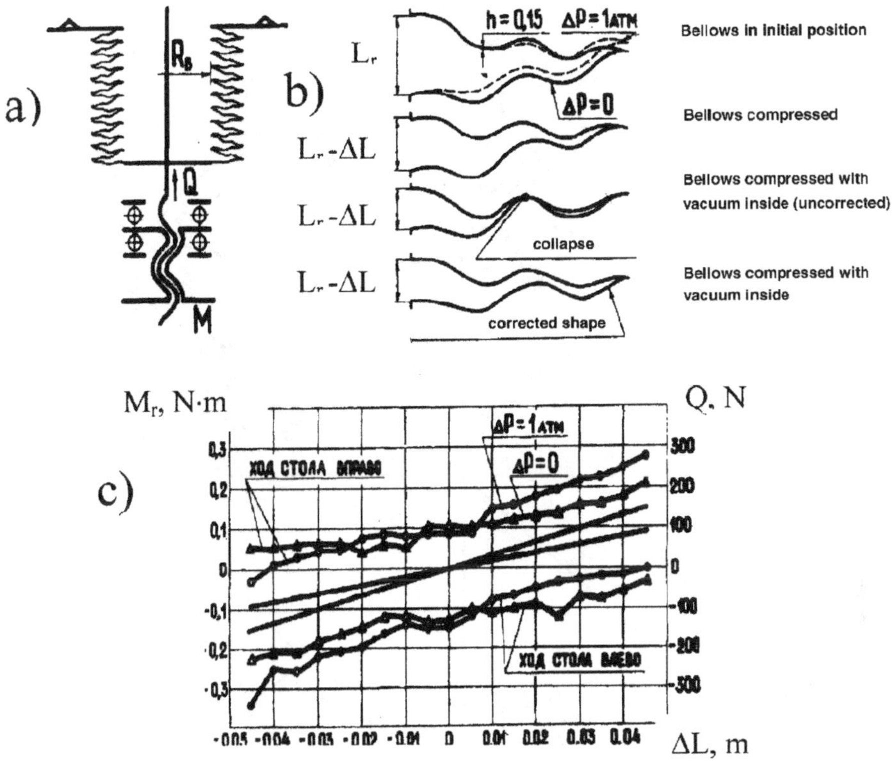

Fig. 5.16 The scheme of positioning error forming as a function of hermetic bellow pressing in the nut-screw drive. (a) Kinematic scheme of the drive, (b) scheme of the bellow deformation with different form of the waves, (c) the resistant force moment of the bellow without pressure difference ($\Delta P = 0$). The other form is a result of atmosphere pressure difference ($\Delta P = 1$ atm).

The features of ball-ring retainer system were noted in some works by Spitsin and Ivanov [22, 26], but only Yurkov with co-authors[2] gave a detailed explanation of this problem.

When the load on the ball bearing is "pure" radial there is a zone of the retainer, where the balls are unloaded and can move freely inside the retainer, thus they cannot be jammed.

When the load on the ball bearing is axial or combined all balls are squeezed between the inner and outer rings; the fastest and the slowest balls come to the opposite insides of the retainer housing, press to the walls and the system ball–ring–retainer is jammed. For continuation of the ball bearing rotation it is necessary to overcome the friction forces between ball–ring or between the ball and the retainer.

[2] Deulin E.A., Papko V.M., Yurkov U.V., Influence of the vacuum ball bearing reliability on the output of atomized technological equipment, in *Abstracts of the Conference USSR BMSTU*, Moscow, 1979, pp. 38–44 [in Russian].

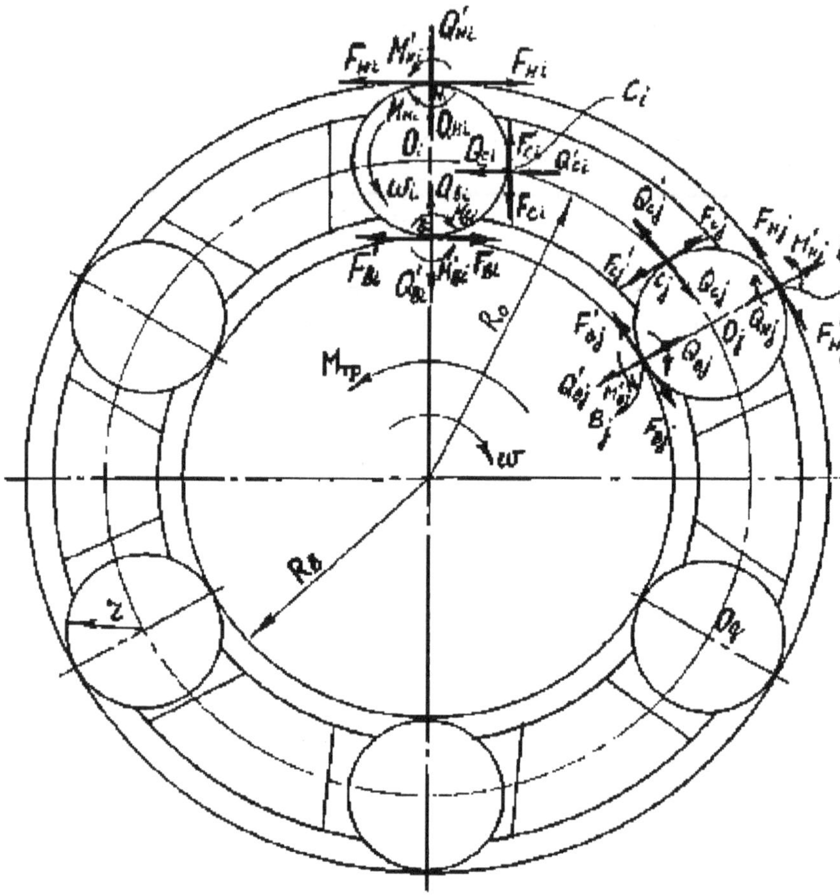

Fig. 5.17 The flat model of the forces action between elements in ball bearing.

Thus, resistance torque of the ball bearing grows till the jammed ball begins to slide along the rings or along the surface of the housing of a retainer. The arising surfaces have an accidental character till the moment of the sliding beginning.

The scheme of forces interaction inside the ball bearing at the moment of sliding beginning is shown in Figure 5.17. The forces acting on the "leading" ball are denoted by index i, the forces, acting on the "driven" ball are denoted by index j, the forces acting on the "neutral" ball are denoted by index g. The forces, which act on the retainer and on the rings are denoted by a prime. The inertia forces are considered negligible.

The equation for the leading ball (with the center in point O_i) forces balance:

$$\sum M_{O_i}(F_i) = 0; \quad F_{Bi} \cdot r + F_{Hi} \cdot r - F_{Ci} \cdot r - M_{Bi} - M_{Hi} = 0,$$

$$\sum F_{xi} = 0, \quad F_{Bi} - F_{Hi} - Q_{Ci} = 0,$$

$$\sum F_{yi} = 0, \quad Q_B - Q_H - F_{Ci} = 0, \tag{5.42}$$

where F_{Bi}, F_{Hi} are the ball friction forces with inner and outer rings, respectively; F_{Ci} are the friction forces between the ball and retainer; Q_{Bi}, Q_{Hi} are the ball normal reaction forces in contacts with inner and outer rings, respectively; Q_{Ci} are the ball normal reaction forces in contacts with the retainer; r is the ball radius; M_{Bi}, M_{Hi} are the ball friction rolling torques with inner and outer rings, respectively.

For the solution of Equation (5.42) we use the following equations:

$$M_{Bi(Hi)} = K \cdot Q_{Bi(Hi)}, \quad F_{Ci} = Q_{Ci} \cdot f_{Ci}, \tag{5.43}$$

where f_{Ci} is the friction coefficient ball-retainer; K is the rolling friction coefficient ball-retainer. $M_{Bi(Hi)}$ are the ball friction forces with inner and outer rings, respectively; F_{Ci} are the friction forces between the ball and retainer.

The solution is the expressions of leading ball friction forces with inner and outer rings:

$$F_{Bi} = 0.5 \cdot Q_{Ci} \cdot (f_{Ci} + 1) + (Q_{Bi} + Q_{Hi}) \cdot K/(2 \cdot r),$$

$$F_{Hi} = 0.5 \cdot Q_{Ci} \cdot (f_{Ci} - 1) + (Q_{Bi} + Q_{Hi}) \cdot K/(2 \cdot r). \tag{5.44}$$

The values Q_{Bi} and Q_{Hi} can be found from the statically indeterminable equation (5.43) using compatibility of transference condition, where force F_{Ci} is used.

The torque in a real ball bearing according presented method of calculation can be estimated as follows:

$$M_{mp} \approx R_B \cdot f_c \sum_{i=1}^{z} Q_{ci} + (R_B + \cos\alpha)k \cdot P_{oc}/\sin\alpha + z \cdot M_{ib} \cdot \sin\alpha, \tag{5.45}$$

where P_{oc} is the axial load of the ball bearing.

The maximum torque we obtain in the case when the sum in Equation (5.45) is maximum; the minimum torque is when only one ball lead the retainer. This minimum torque can be calculated as

$$M_{mp} = (R_B/r + \cos\alpha)kP_{oc}/\sin\alpha + z \cdot M_{ib} \cdot \sin\alpha. \tag{5.46}$$

The results of theoretical calculation according to this model and the experimental results for comparison are presented in Figure 5.18. The described accidental fluctuation of resistance torque of ball bearing causes the appearing of the new elastic error component, which can be estimated:

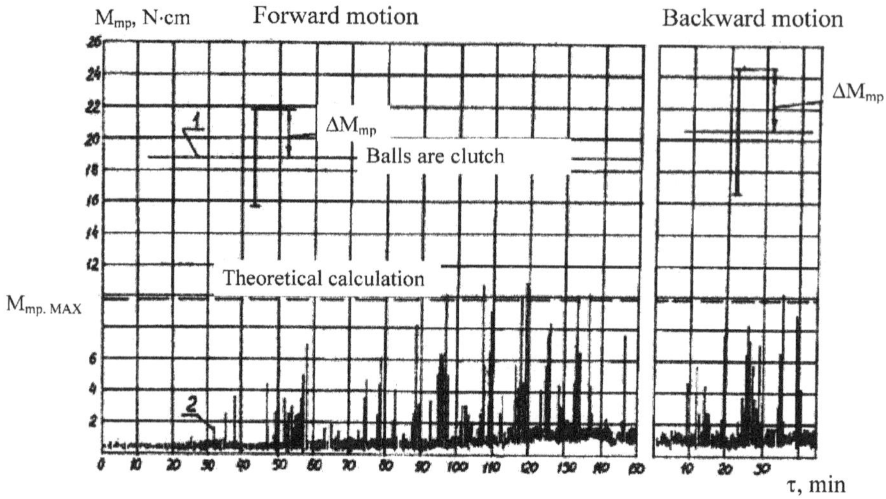

Fig. 5.18 The results of theoretical calculations according to the above presented equations and the experimental results for comparison ($P = 10^{-5}$ Pa): 1 – sliding $P_{ax} = 10$ N; 2 – rolling $P_{ax} = 30$ N.

$$\delta_8 = \delta\varphi_c = \sum_{k=1}^{c}\sum_{i=1}^{k} M_{mp} \cdot \varepsilon_i / J_{i1}. \quad (5.47)$$

This error component is accidental in its form.

• *Error component of planetary nut-screw drives due to the own resistance torque fluctuation*

Research of vacuum planetary nut-screw drives shows that they have own resistance torque fluctuation, the ones in the ball bearing are similar. The resistance torque fluctuation of these drives reach 200 to 300% of nominal resistance torque.

The theoretical analysis of the work principle of planetary nut-screw drives shows, that some "fast" planetary rollers push the retainer of nut-screw drive ("leading" rollers), the "slow" rollers ("driven" rollers) are retarded by the retainer like it takes place in a ball bearing. It occurs as a result of accidental warp of roller axis. The appearing accidental forces R_c between the retainer and rollers necks create the additional resistance torque moment which has an accidental character. The scheme of forces action in the researched planetary nut-screw drive is shown in Figure 5.19.

On the analogy of a ball bearing research method, the equations of forces and moments balance have been created and maximum and minimum resistance torque have been calculated:

$$M_{mp,\max} = \sum F_{fb(\Gamma)} r_{B(\Gamma)} + R_c r_u + M_{mpB(\Gamma)}, \quad (5.48)$$

$$M_{mp,\min} = \sum R_{fb(\Gamma)} k_{B(\Gamma)} + P_{oc}\tan(\alpha)(k_B + k_\Gamma), \quad (5.49)$$

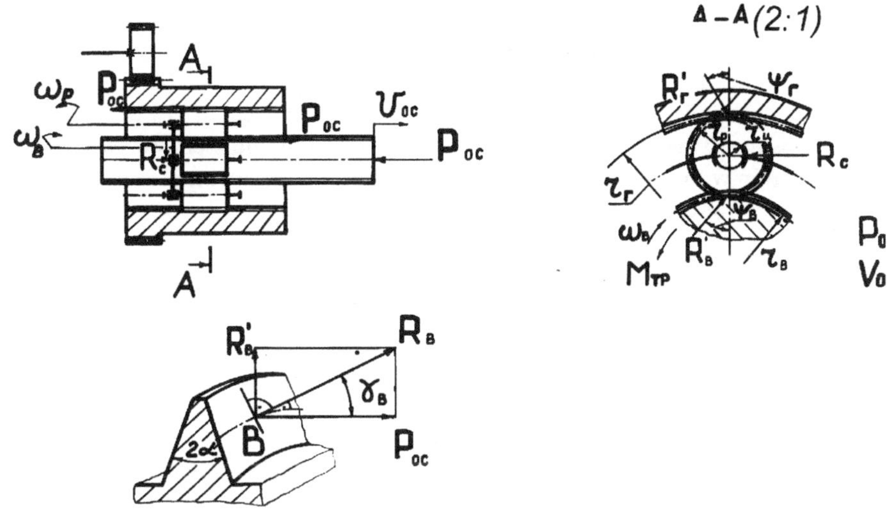

Fig. 5.19 The scheme of forces action the planetary nut-screw drive.

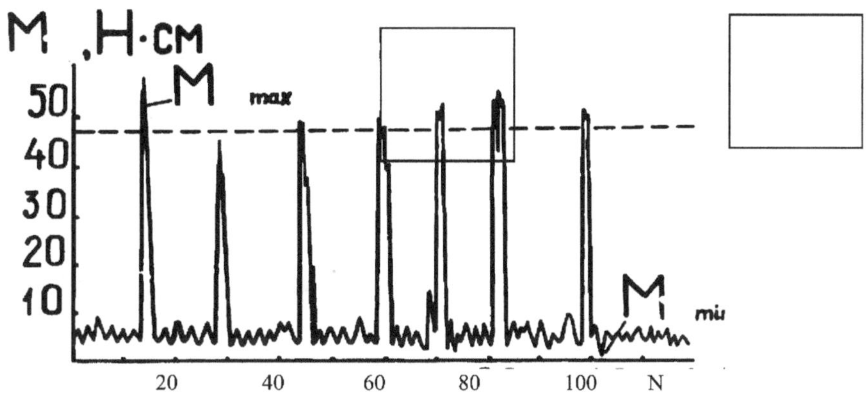

Fig. 5.20 The theoretical (1) experimental (2) friction torques in planetary nut-screw drives.

where F_{fB}, $F_{f\Gamma}$ are the friction forces which act on the screw and nut, respectively; r_B, r_Γ are the radii of the screw and nut, respectively; R_c is the friction force which acts between the retainer and rollers neck; r_u is the radius of the neck of the rollers, which acts with the retainer; M_{mpB}, $M_{mp\Gamma}$ are the moments of friction force in the couplings screw-roller and nut-roller, respectively; R_B, R_Γ are the normal forces in the separator couplings with nut and with screw, respectively; k_B, k_Γ rolling friction coefficient of the rolls with screw and nut, respectively; P_{oc} is the axial load; α is the angle of thread.

As we can see, the resistance torque depend on the axial load and on the thread parameters. Figure 5.20 shows the experimentally obtained diagram of resistance torque variation of planetary nut-screw drives in time also as the value of the resistance torque calculated according to Equation (5.48) for comparison. As we can see the resistance torque fluctuation has the accidental character and the value of this fluctuation is high because of high friction coefficients in vacuum.

The considered resistance torque fluctuation of planetary nut-screw drives in time in combination with small rigidity of the kinematic chain elements creates the additional elastic error component:

$$\delta_8 = \delta\varphi_c = \sum_{k=1}^{c}\sum_{i=1}^{k} M_{mp} * \varepsilon_i / J_{i1}. \tag{5.50}$$

• *Error component due to the vacuum influence on the resistance forces and friction forces*
Error component due to the vacuum influence on the increase of friction forces or on appearing of additional resistance forces has accidental or linear character and can be expressed as

$$\delta_6 = \delta\varphi_c(\delta\varphi_l) = \sum_{k=1}^{c}\sum_{i=1}^{k} F_{TK} * \varepsilon_i / J_{i1}, \tag{5.51}$$

where F_{TK} are the forces caused by friction or by pressure difference which deform the kinematic chain of the drive; k is the number of the factors taken into account, $k = 1 \ldots C$; i is the number of the kinematic chain elements, $i = 1 \ldots n$; $\delta\varphi_c$, $\delta\varphi_l$ are the errors taken into account as accidental or linear.

The error determined by Equation (5.51) can be considered as a constant at stable friction coefficient or at a constant pressures difference but also as accidental variable.

The error due to instability of friction forces is determined as a accidental value:

$$\delta_7 = \delta\varphi_c = \sum_{k=1}^{c}\sum_{i=1}^{k} (1 \ldots 2) F_{TK} \cdot \varepsilon_i / J_{i1}, \tag{5.52}$$

where $(1 \ldots 2)F_{TK}$ is the range of a friction force variation.

The accidental factors mentioned – the accidental variation of friction coefficient, the accidental variation of resistance torque, the accidental value of the difference between static and dynamic friction coefficients – are simulated as accidental components of the total error of positioning.

• *Error component due to the outgassing baking*
The ultrahigh vacuum (UHV) mechanisms must be periodically baked at temperatures of 200 to 400°C to ensure low outgassing rate. The periodical baking at high temperatures causes the residual deformations of the drive elements to appear. This transforms the initial balance of errors and makes it worse. These deformations are the result of the heat deformations and creep deformations of basic details. A num-

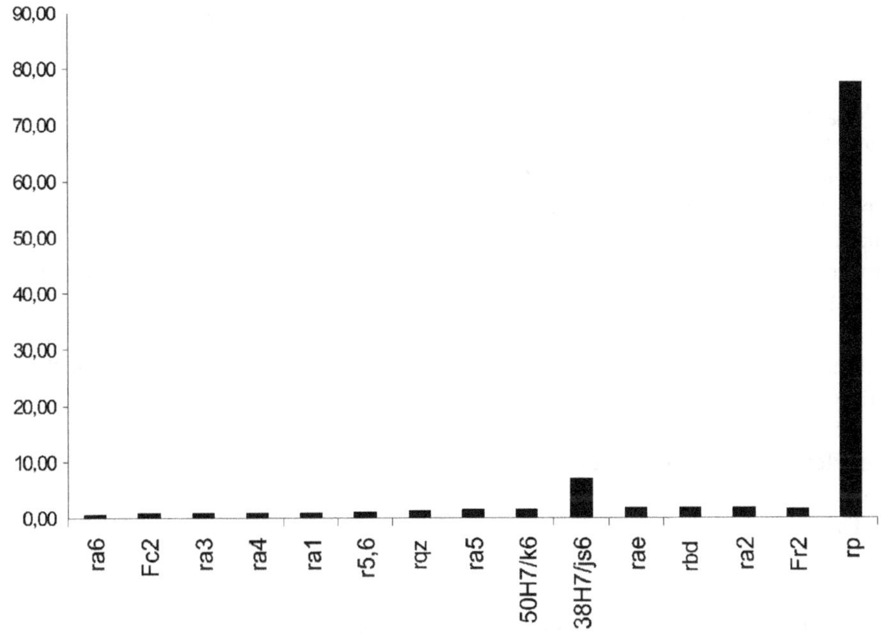

Fig. 5.21 The part (in percentage) of error components of hermetic gears planetary feedthrough, where ra_i – radial inclinication of i number ball-bearing traces, r_{bd}, r_{ae}, r_{gz}, r_{56} – radial inclinication of the base surfaces titled, also as of ball-bearing surfaces BD, AE, GZ of ball-bearing base surfaces 5, 6; F_{r2} – inclination of a gear size; F_{c2} – inclination of a gear size; 50H7/k6, 38H7/js6 – the gaps in the corresponding corrections; r_p – the corresponding axis inclinations of the surfaces B, D as a result of the element movement (compare with Figure 5.9).

ber of publications [11–15, 20, 21, 25] is dedicated to this problem, however the problem remains till nowadays.

It was shown in [23] that after the baking of the bellows-sealed rotary-motion feedthrough, creep deformation of the bellow appears. These deformations depend on the temperature duration of heating and initial elastic deformation of the bellow. At the same time the residual deformations appear under stresses much below the yield stress at the baking temperature.

The research of the typical body elements of UHV feedthroughs baked at 200–450°C in the loaded state (bended state) shows that the increasing residual deformation has a complicated character and finally is about 100–140 microns. The vector of the residual deformation is directed in the direction of the bending force. This deformation must be taken into account in the calculation of the total error of positioning. The value of the deformation has a large dispersion and does not depend on temperature in the used range of temperatures (200–450°C), which points to the creep saturation.

Comparison of the total error components shows that the residual deformations due to baking are comparable with the technological errors and mounting errors.

Figure 5.21 shows the balance of the errors of a gear planetary rotary-motion feedthrough (see Figure 5.5). It can be seen that after one heating of the body of this feedthrough (length of body is 110 mm, diameter is 40 mm, wall thickness is 2 mm, baking temperature is 250–450°C, time of baking is 4–29 hours) the residual deformation was 6–38 microns.

In addition to the direct influence of this deformation to the total error of the drive, the creep residual deformations cause its own resistance force to increase. That leads to additional elastic error. To calculate these resistance forces it is necessary to analyze the initial errors of size of the basic elements of the drive, together with the creep residual deformations.

Let us consider the balance of the total error components including heat creep residual deformations of basic elements of a gear planetary rotary-motion feedthrough (Figure 5.5).

When the instant value of radial displacement of gear ε increases the value of abducted gap b as the interference of gearing begins, that causes its own resistance torque to increase.

The total resistance torque reduced to leading shaft can be presented as a sum of the following components: the reduced resistance moment of external load M_{U1}, the reduced resistance moment of elastic forces due to elastic deformations M_E, the reduced resistance moment of the deformation of sealing elements (sealing thin-wall tube, flexible gear, sealing bellow) M_H, the reduced resistance moment of friction forces in gear coupling M_P, the reduced resistance moment of ball bearings M_{bb}:

$$M_{TP} = M_{U1} + M_E + M_H + M_P + M_{bb}. \tag{5.53}$$

So, the computer simulation of the total error takes into account the value of abducted gap b, the instant value of radial displacement of gear ε, the reduced value of the drive rigidity J_I and generalized coordinate (in a role of a generalized coordinate, the degree of leading shaft rotation α is taken):

$$M = f(b, \varepsilon, J_I, \alpha, \dots). \tag{5.54}$$

The components M_E and M_H caused by additional elastic forces are the dissipative factors and have the symmetrical (in respect to zero) cycle of variation, which can be positive or negative. The components M_P and M_{bb} are non-dissipative factors.

The value of total resistance torque calculated according to Equation (5.54) allows the estimation of the average resistance torque. The dispersion of the resistance torque can be calculated using the following equation:

$$\sigma(M_{Tp}) = \sqrt{\left(\frac{\partial M_{Tp}}{\partial \varepsilon}\right)^2 \sigma^2(\varepsilon) + \left(\frac{\partial M_{Tp}}{\partial b}\right)^2 \sigma^2(b)}. \tag{5.55}$$

The calculated numbers allow the estimation of the additional error components using Equations (5.47) and (5.52), as well as the estimation of the parametrical reliability on "precision" (error of positioning) parameter according to Equations (5.5) and (5.9).

• *Time law error of positioning*
The dynamic component of the positioning error has a comparable and sometimes dominating value when the error of positioning is determined in a short time period after a mechanism abrasion or after variation of the parameters of the movement. It can become dominating because of the small rigidity of vacuum mechanisms.

The form of the dynamic component is different from other types of error components (linear, cyclical, accidental or constant components) and can be expressed as

$$\delta_{\partial uh} = A_0 e^{-nt} \sin(kt + \varphi_0), \tag{5.56}$$

where A_0 is the initial deformation from inertia load; n is the damping coefficient; t is the current time; φ_0 is the initial phase of vibration; k is the vibration frequency.

The parameters of Equation (5.56) can be determined from the handbooks on elastic mechanics.[3]

The above-mentioned error component is typical for the special frictionless vacuum mechanisms based on controlled deformations (see Chapter 9).

5.4 Summarizing the Components of different Types and Forms

As was shown above, the variation in the type of error components may appear in different forms: linear, cyclical, accidental, dynamic form of variation as is shown in Figure 5.22.

Kinematic error components have a linear or harmonic form. Elastic error components have a linear or harmonic form in the case of linear or harmonic deformation of the elastic element (e.g., bellow, spring, flexible gear), but at the same time, elastic error components have an accidental form in the case of non-stable friction forces or accidental mechanical interaction of mechanical elements of the drive (ball bearing elements, planetary nut-screw drive elements). Dynamic components have a short time damping.

So, for the total error calculation all components should be expressed in one of the mentioned forms. Then, during the computer simulation, they are summarized according to the equation, which joins the position of the output element of the mechanism in coordinate φ_0 with parameters of the mechanism:

---

[3] Biderman V.L., The mechanics of thin-wall designs, *Mechanical Engineering*, 1977, p. 488 [in Russian].

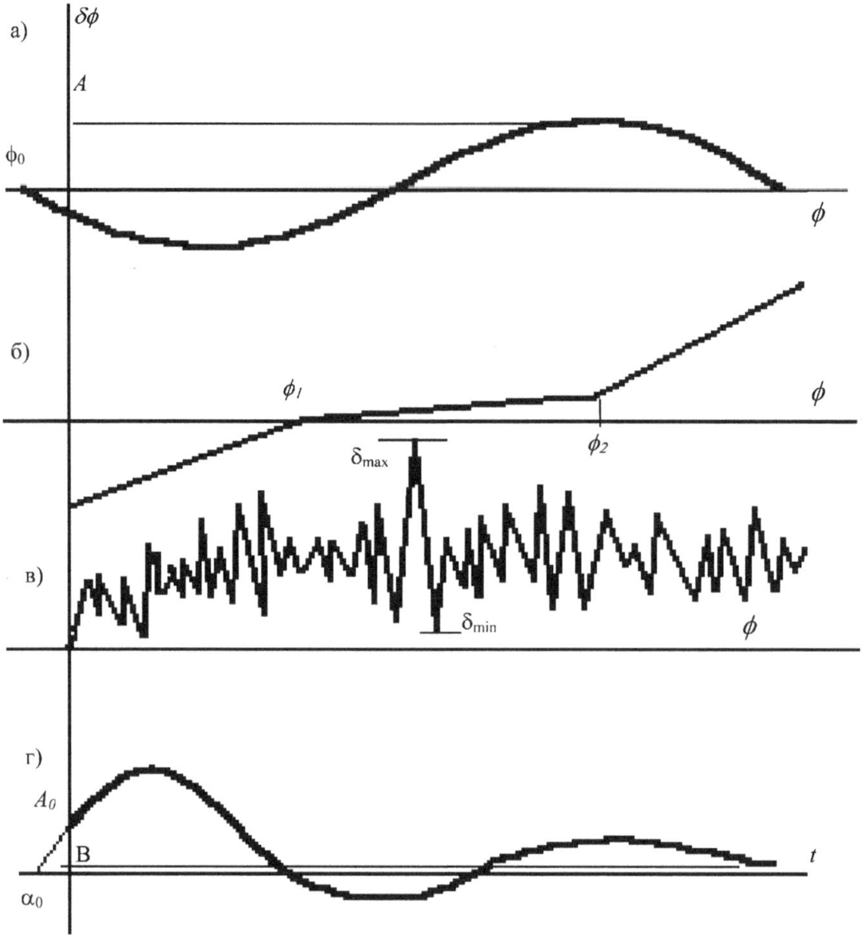

Fig. 5.22 The main forms and components of positioning errors. (a) Harmonic component, (b) component, (c) accidental component, (d) dynamic component.

$$\begin{aligned} d\varphi &= \varphi_0 + d\varphi_D = \varphi_0 + \Sigma\delta\varphi_\Gamma + \Sigma\delta\varphi_c + \Sigma\delta\varphi_L + \Sigma\Delta\varphi_D \\ &= \phi\phi_0 + \Sigma r_i k_m \varepsilon_i \sin((n_i\phi + \phi_{Hi})/\varepsilon_i') + \Sigma\delta\varphi_u\varepsilon_i + \Sigma j_{\varphi\Sigma}\varepsilon_i \\ &\quad + \Sigma j_{\varphi u}\varepsilon_i + P/J_{n1} + \Sigma\Sigma F_{TK}\cdot\varepsilon_i/J_{i1} + (1\ldots2)F_{TK}\cdot\varepsilon_i/J_{i1} + \\ &\quad + \Sigma\Sigma\delta F_{TK}\cdot\varepsilon_i/J_{i1} + +\Sigma\Sigma(dE_k/d(\varphi/\varepsilon_i))/(J_{i1}\varepsilon_i) + \\ &\quad + A_0 e^{-nt}\sin(kt+\varphi_0). \end{aligned} \tag{5.57}$$

In the majority of cases a transmission coefficient of errors ε_i, ε_{in}, are inversely depend on transmission coefficient u_i, u_{in} of a corresponding part of kinematic part of a drive.

The drive sections which work in vacuum have the parameters of Equation (5.57) (which are instable) and must be taken into account as accidental values. That is why the final position of the researched object is also an accidental value. It can be expressed by the value $\sigma(\delta\varphi_\partial)$, which can be used for calculation of the reliability of the drive:

$$\sigma(\delta\varphi_\partial) = \sqrt{\sum_{j=1}^{\varphi} \left[\frac{\partial(\delta\varphi_\partial)}{\partial(\delta x_j)} \right]^2 \sigma^2(x_j)}, \tag{5.58}$$

where $\delta\varphi_\partial$ is the number of the parameters which are considered as accidental values and which have an influence on the stability of the object transference; $\partial(\delta\varphi_\partial)/\partial x_j$ is the partial derivative of a function on x_j parameter ($x_j = r_i$, j_i, and so on), this is considered to be an accidental value; $\sigma(x_j)$ is the standard deviation of x_j parameter.

The particular derivative analysis shows that different factors, such as an error of entrance, initial error of dead speed inflexibility must not bear influence on the error of the drive. These factors have a great influence only at small reduction of the mechanism and in the final section of the mechanism.

The vacuum movement feedthroughs usually are the ultimate transmission elements of the drive. The aim is to exclude the influence of vacuum mechanical elements on the residual atmosphere of the vacuum chamber. That is why the parameters of the vacuum feedthroughs (which enter into Equation (5.57): kinematic errors, resistance forces, recoil, inflexibility) have a great effect on the precision of the drive.

Calculation of the mean square error $\sigma(\delta\varphi_\partial)$, is necessary for parameter reliability of vacuum equipment calculation and it has a number of specific features.

The first feature of this equipment is a large dispersion and large values of resistance forces. The second feature of this equipment is the presence of specific pressurizing elements: steel bellows, thin-wall rubber tubes, cuffs, which vary the rigidity of the drive and create varying (not linear) resistance forces of the drive. The third feature of this equipment is the pressures difference which appears on the motion feedthroughs. These feedthroughs have an influence on the resistance force F_{TK} and the drive rigidity j_i. The fourth feature of this equipment which have an influence on the outgassing rate and on the equipment cleanness is complicated kinematics of the motion feedthroughs which detects the kinematic error $\delta\varphi_\Sigma$ of the last kinematic part of the total kinematic chain.

The error of computer calculation based on Equation (5.57) in the area of φ variation with the step variation $\delta\varphi$ allows one to reach the following probable parameters of the positioning error:

1. The true total error of transference can be considered as a function of the rotation φ angle calculated using the computer calculation as a function of $\delta\varphi$ in the result of computer simulation.

2. The mathematic expression of the error of positioning for the set of the mechanisms assembled in accordance with the work drafts can be expressed as

$$M([\delta_\varphi]) = \frac{\sum_{k=1}^{N \cdot N_j} [\delta_{\varphi\partial}]}{N \cdot N_j}, \tag{5.59}$$

where N_j is the number of the calculation cycles (number of the graph points) of a mechanism in the calculated interval of the rotation angle φ; N is the number of the mechanisms in the considered set.
3. The standard deviation of the transference for the considered set of the mechanisms:

$$M([\delta_\varphi]) = \sqrt{\sum_{k=1}^{N \cdot N_j} ([\delta_{\varphi\partial i}] - M([\delta_{\varphi\partial}])^2 / N \cdot N_j)}. \tag{5.60}$$

4. The probability of not exceeding the allowable error of transference by the real error of transference in the observed set of the mechanisms:

$$P(\delta_{\varphi D}) = 1 - m/N_j, \tag{5.61}$$

where m is the number of exceeding of the calculated allowable error $\delta_{\varphi D}$ by the real error $\delta\varphi_{\Sigma Di}$.
5. The probability of not exceeding the allowable error for the set of the mechanisms of good quality:

$$P(\delta\varphi_D) = 1 - m_\Sigma / N_j * N, \tag{5.62}$$

where m is the total number exceeding permissible error in the considered set of the mechanisms.
6. The probability of not exceeding the permissible error as a function of permissible level of the error.
7. The comparative balance of different frequency errors for the considered set of the drives:

$$K_j = \frac{\sum_j^N A_j}{N \cdot M([\delta\varphi_{\Sigma\partial}])} \cdot 100\%. \tag{5.63}$$

8. The absolute balance of different frequency errors for the considered set of the drives:

$$K_{aj} = \frac{\sum_l^N [M([\delta\varphi_{\Sigma\partial}]) - M([\delta\varphi_{\Sigma\partial}^i])]}{N \cdot M([\delta\varphi_{\Sigma\partial}])}, \tag{5.64}$$

where $M([\delta\varphi_i])$ is the average distribution of transference error for the considered set of the drives without the i component of the error.

The last two probable parameters of the positioning error help to "analytically" find the dominating components of the summarized error and to find the constructive designs with the least amount of errors.

5.5 Correlation of Total Error of the Mechanisms with Economic Parameters

It was shown in the beginning of this chapter that it is possible to use two different ways of error decreasing:

- loop-controlled system using a sensor to detect the positioning error,
- open system which uses special technological ways to diminish the drive error.

Let us consider the efficiency of the production costs for different ways of the kinematic error decreasing for the mechanism based on harmonic rotary feedthrough (see Chapter 7).

In the described method, the method of total error estimation in the design process is discussed. It was shown that the total error of the drive depends mainly on the precision of the vacuum motion feedthrough.

The theoretical analysis of the assembling process of vacuum feedthroughs and the real results of the experience show, that the total drive error depends on the precision of main elements and this precision depends on several main manufacturing operations.

These main elements of the harmonic rotary feedthrough are: body, leading gear, leading shaft, thin-wall gear. When the precision of these details increases, then additional technological operations are necessary and that causes the manufacturing costs to go up.

In the example considered, the body of a harmonic rotary feedthrough has a precision of its main elements of about 7 to 8 (ISO standard of precision quality) and the time of machining is 9.8 min.

The additional operation of machining allows one to achieve the precision of main elements about 6 to 7 (ISO precision quality scale). This additional operation is fine grinding. This operation increases the operating time of the machine to 12.6 min.

If the additional operation of machining achieves the precision of main elements about 5 to 6 (ISO-scale), the additional operation should be lapped finishing. This operation increases the machining time to 153 min.

Machining of the leading shaft machining ensures the precision of this component at 7 to 8 (ISO-scale) with machining time 19.8 min. The additional operation to obtain precision at 5 to 6 (ISO-scale) uses the lapped finishing operation and special hydro-plastic base device which increases the machining time to 23.3 min.

Machining operation of leading gear corresponding to 7 (ISO-scale) includes nine machining operations with the machining time of 32.1 min.

When special technological devices are used to obtain 5 (ISO): special base details and cutting instrument, the number of machining operations increases to 10 and the machining time increases to 66.2 min.

The total machining time of main elements of the drive can be expressed as a function [1, 2] of the precision of main drive elements:

$$T = K/\delta\varphi, \tag{5.65}$$

where T is the total machining time of main elements of the feedthrough; $\delta\varphi$ is the feedthrough precision (angular errors); K is the coefficient.

The main parameters of the process are: working vacuum $5 \cdot 10^{-8}$ Pa, production yield is 2000 feedthrough units per year, the costs of one unit is 10^5 Russian roubles per unit. The required precision of a unit is four degrees. The running costs can be expressed by

$$C_{\exp} = C_{0\exp}/h_{\text{pos}}, \tag{5.66}$$

where h_{pos} is the coefficient of good quality details of output. This coefficient can be determined as shown above in Section 5.4.

As was shown above, the drive precision has an influence on the running costs. It shows that calculation of the running costs and determination of the economically profitable drive variant can be calculated according to [3]. The economical efficiency C_{ee} can be calculated:

$$C_{ee} = C_1 - C_2 = (C_{e1} + E_H K_1) - (C_{e2} + E_H K_2), \tag{5.67}$$

where C_1, C_2 is the ideal total running costs for ideal and for calculated drive variants. C_{e1} and C_{e2} are the real running costs for ideal and for calculated drive variants. K_1, K_2 are the corresponding running costs. K_1 is the price for a typical drive $K_2 = K_0 C_{ff}$, where C_{ff} is the prime cost of the feedthrough, K_0 is the coefficient of modernization price.

In the case of required precision 3° to 7°, it is economically profitable to increase the precision of main elements using modern technological methods to 4 to 5 (ISO).

The material considered shows that there is an optimal precision of a non-loop-controlled drive which is equal to $\pm 1°$ for the considered material. We also know that the limit of the drive error $\Delta\varphi = [\delta\varphi] = 4°$ while the limit that the drive error could be decreased is $\pm 1°$ to 2°. The example also shows that further decreasing drive errors will require the use a loop-controlled drive.

References

1. Pronikov, A.S., *Precision and Reliability of Digitally-Controlled Machines*, Machinostroenie, Moscow, 1982, 250 pp. [in Russian].
2. Dunaev P.F., The rational methods of tolerances calculation, *Machine Tools and Instrument* **6**, 1952, 32–47 [in Russian].
3. Baranov G.G., About the choosing of the tolerances to ensure the demanded precision and the price of the mechanism, in *Institute Mashinovedemia Proceedings*, Seminar of Precision in Machinebuilding, Vol. 11, Academy of USSR, Moscow, 1957, pp. 86–94 [in Russian].
4. Rasumov I.M., *Economics in the Machinebuilding of USSR*, The "High School" Edition, Moscow, 1982, 452 pp. [in Russian].
5. Alexandrova A.T., The theoretic base of optimal vacuum mechanisms design, Thesis of Scientific Doctoral Research, Scientific direction number 05.02.12, Moscow, 1979, 46 pp.
6. Satarov, G.K., Deulin E.A., Vasichev B.I., The electron beam lithography equipment productivity, Electronics revision, 7 Issue "Technology organization and equipment", 1986, v. 14 (1219), TSNII, Moscow, "Electronics", 44 pp. [in Russian].
7. Borodachev N.A., *The Production Precision*, Moscow, 1973, 372 pp.

8. USSR standard 1643-81, The base of design base. Cylinder gears: Tolerances. Moscow Tolerances, Moscow, 1985, 69 pp. [in Russian].
9. USSR Standard 21098-82, Kinematic chains. The methods of precision calculation. Standard, 1984, 26 pp. [in Russian].
10. Deulin E.A., Mednikov M.I., Papko V.M., Vacuum mechanisms design and calculation/"Zaocny Institute ITR", "Machinostroenie", Moscow, 1986, 80 pp. [in Russian].
11. Deulin E.A., Pavlov A.N., The planetary vacuum feedthroughs precision increasing, High school edition, *Machinebuilding* **10**, 1982, 37–42 [in Russian].
12. Zdanovith T.K., The heat and mechanic action on special clean stainless steel to increase its work properties, Scientific Work Thesis, Moscow, 1985, 16 pp. [in Russian].
13. Zaripova R.G., Kaibishev A., Salishev, G.A., The properties and structure variation of stainless steel under action of crystals, *Metals Research and Heat Action* **9**, 1993, 19–22 [in Russian].
14. Ilichev V.Y., Startsev V.I., Shapovalov I.A., The stainless steels creep at low temperatures, in *Proceedings Physical Process of Plastic Deformation at Low Temperature*, Naukova dumka, Kiev USSR, 1974, pp. 358–361 [in Russian].
15. Kennedy A.Dj., *Creep and tired Process in Metals* (English original), USSR Ed. "Metallurgia", Moscow, 1965, 312 pp. [in Russian].
16. Koval V.A., Low temperature creep of some metals, Thesis of Scientific Doctoral Research, Charkov, Ukrain USSR, 1975, 16 pp. [in Russian].
17. Kojevnikov V.P., Korotkova L.V., The methods of precision calculation in electronics industry, *Electronics Revision* **8**(2) (359), TSNI "Electronics", 1986 Moscow, 42 pp. [in Russian].
18. Kutsokon V.A. et al., *The Probability Theory at the Mechanism Design Usage*, USSR Leningrad, "Mashinostroenie" Edition, 1971, 260 pp [in Russian].
19. Kutsokon V.A., *The Precision of Kinematik Chains of Mechanisms*, USSR Leningrad, "Machinostroenie", 1980, 221 pp. [in Russian].
20. Rabotnov Y.N., *The Creep Process in mechanical Devices*, Nayka, Moscow, 1966, 752 pp. [in Russian].
21. Volchkevith L.I., Kamyshny N.I., Deulin E.A., *Handbook for Students*, Bauman Moscow State Technical University, Moscow, 1985, 60 pp. [in Russian].
22. Starostin V.F., The ball movement in the ball bearing, in *Proceedings of Soviet Union Bearing Research Institute (VNIINP)*, Vol. 1 (45), pp. 42–53 [in Russian].
23. Akchmadiev D.R., Deulin E.A., About reliability of baking ultra high vacuum mechanisms, in *Proceedings 12th International Vacuum Congress*, Book of Abstracts, The Hague, 1992, p. 88.
24. Deulin E.A., Usov A.B., Marusov V.A., et al., *IVC-12: Book of Abstracts*, The Hague, 1992, p. 80.
25. Garofalo F., *Fundamentals of Creep and Creep Rupture in Metals*, MacMillan, New York, 1965.
26. Sinodeev I.V., Deulin E.A., Yurkov Y.V., Research of the resistance forces in vacuum planetary rotation motion feedthroughs, Computer Edition (VNIITEMR deponent No. 3888.-18.10.85), Moscow, 1985 [in Russian].
27. Fursak F.I., The small inertia of harmonic rotary drive research, Thesis of Scientific Ddoctoral Research, Moscow, 1973, 16 pp. [in Russian].
28. Firsaev A.F., The design of the harmonic rotation gear drive of the computer based lathes, Thesis of Scientific Doctoral Research, Moscow, 1983, 16 pp. [in Russian].
29. Chernianski P.M., *The Metal Cutting Lathes Rigidity*, Bauman Moscow State Technical University, Moscow, 1969, 80 pp. [in Russian].
30. Chernova L.S. et al., The precision of different harmonic rotation drives research, in *Proceedings of Leningrad Conference*, Leningrad, 1968, pp. 110–115 [in Russian].
31. Chernianski P.M., The rigidity of statically varied system research, *Machine Building* **8**, 1981, 127–130 [in Russian].
32. Chernianski P.M., The scientific base of high precision of the mechanically loaded lathes, Thesis of Scientific Doctoral Research, Moscow, 1986, 32 pp. [in Russian].
33. Shuvalov S.A. et al., The parameters of precision and rigidity of harmonic rotary drive, *Machine Building* **6**, 1970, 56–62 [in Russian].

34. Push V.E., The dynamic of lathes, *Mechanical Engineering*, 1968, 560 [in Russian].
35. Deulin, E.A., *Mechanics and Physics of Precise Vacuum Mechanisms*, Vol. 1, Vladimir State University, 2001, 176 pp. [in Russian].

Chapter 6
Vacuum Mechanisms of Nanoscale Precision

The principles of design of vacuum mechanims of nanoscale precision are presented. Physical basics of the mechanisms nanoscale precision are discussed. Vacuum multicoordinate drives and manipulators are also shown.

The analysis of vacuum technological equipment and vacuum research equipment shows that the most strict requirements for the object transference are in electron beam micro lithography equipment. For example, electron beam micro lithography installations with work pressure $p = 10^{-5}$ Pa must ensure the positioning error is below 0.5–1.0 mic. The total error of instrument (and coordinate table) [1] positioning in this installation is below 100 nanometers. Even higher requirements are in X-ray lithography installations which use synchrotron source as a work beam [2]. In the process of sample transference this equipment requires that the silicon water inclination from the base X-ray sample should be below 20 nanometers. These high requirements are caused by the work beam inclination. In the X-ray installations, the work beam is not inclined by the electric field as in the electron beam lithography. In new research equipments: in the ultrahigh vacuum scanning tunneling microscopy installations, in atomic force microscopy, the error of the instrument transference must be less than 0.1 nanometer (at a work pressure $p \leq 5 \cdot 10^{-8}$ Pa) [3].

The other types of research equipment, for example ultralarge segmented astronomy telescopes equipped with an adaptive optical system, requires small positioning error less than 50 nanometers. This system also has special requirements on cleaniness and absence of debris (products of wear) [4]. This equipment requires also very quick action of the drive (the time of action is less than 10^{-1} micro-second). It is related with the necessity of high productivity and multi-step transference of

E.A. Deulin et al., *Mechanics and Physics of Precise Vacuum Mechanisms*, Fluid Mechanics and Its Applications 91, DOI 10.1007/978-90-481-2520-3_6,

the sample with high precision. The quick action of the drive can be achieved with the high stability of the motion of the drive: the equipment requires short time decreasing of the transference process as well as high speed of sample transfer [5]. To ensure nanometer precision in combination with millisecond quick-action of the drive, it is necessary to use the correct type of the drive and to have a system with computer control. For this task it is necessary to establish the principles of transference of ultrahigh precision of vacuum mechanisms.

6.1 The Principles of Nanometer Precision of Vacuum Mechanisms

It was shown in Chapter 5 that only loop-controlled vacuum mechanisms are able to limit the number of positioning errors. It was shown [5] that the total positioning error of loop-controlled drive which works in a stable environment can be expressed by Equation (5.10):

$$\delta\varphi_\Sigma = \delta_V + \delta_N + \delta_{CT},$$

where δ_V is the error of driving action variation (this error is determined by the drive structure and by the drive control system); δ_{CT} is the static error of the drive (this error is limited by the static friction forces in the drive elements and by the measuring system).

Let us consider the total error $\delta\varphi_\Sigma$ components of the drives of nanometer scale. The error of driving action variation δ_V can be expressed by the following equation [5]:

$$\delta_V = C_0\varphi + C_1\omega + C_2\varepsilon, \tag{6.1}$$

where C_0, C_1, C_2 are the errors coefficients: positioning coefficient, speed coefficient, acceleration coefficient, respectively; φ, ω, ε are the drive coordinate, drive speed, drive acceleration, respectively.

In the drive static control system (at a static level $\nu = 0$) the drive error of driving action variation $\delta_V = C_0\varphi$. The drive static control system which contains one integrant element ($\nu = 1$), the positioning error is equal to zero ($C_0 = 0$) and the δ_V error of driving action variation component depends on dynamic error:

$$\delta_V = C_1\omega = \omega/k_\Pi,$$

where k_Π is the common speed transmission coefficient of automatically controlled drive.

In the case of the static system containing two integrant elements, the positioning and speed errors are equal to zero ($C_0 = C_1 = 0$) and δ_V error of driving action variation depends on the acceleration error $\delta_V = C_2\varepsilon$.

In the presence of disturbing forces (F_{dist}) (technology forces variation, vibration of base forces, atmosphere pressure forces variation after the vacuum chamber opening, energy variation and hydraulic pumping station variation), the error [5]

Table 6.1 Parameters of transference measuring systems.

Technical parameters of transference measuring systems	Photoelectric sensor of linear transference (Model BE-164, ENIMS, Russia)	Inductive sensor of of positioning (WS100D ... WS600D, HBM, Germany)	Industrial laser interferometer (HP5528A, Hewlett Packard, USA)
Range of the transference length, m	1.02	0.6	60
Precision of transference measurement, Nm	500	100	10
Maximum speed of transference, m/min	10	–	18

appears:

$$\delta_N = F_{\text{dist}}/(k_{\text{damp}} i^2 k_a), \tag{6.2}$$

where k_{damp} is the damper coefficient of the drive; i is the coefficient of the drive transmission; k_a is the common amplifier coefficient of open control system of the drive.

When using vibro-isolated base units and thermostatic systems as well as in the case of hydraulic pumping stations of high stability, the error δ_N can be reduced to a small value ($\delta_N = 0$). The "static" error δ_{CT} of loop-controlled drive can be determined by Equation (5.11).

This error is caused by a dead zone of the loop-controlled drive. It can be presented as

$$\delta_{CT} = \delta_c + \delta_L + \delta_I \quad \text{(Equation (5.11))}$$

where δ_c is the error component caused by friction forces and other resistance forces (friction forces, elastic forces and pressure forces of sealing bellows and membranes, etc.); δ_L is the error component caused by recoil (gap in a kinematic chains); δ_I is the instrumental error component which is determined by the accuracy rating of a sensor as well as by a zero drift of the amplifier.

Precise vacuum drive must consist of parts (motor, transmission, pilots) without recoil elements. The recoil decreasing in kinematic chains of transmission leads to an increase of the friction force and an increase of the error component δ_c. The use of hydraulic or piezo electric drive elements helps to remove the recoil component ($\delta_L = 0$).

The different types of measuring systems: raster grids, inductive sensors, laser interferometers are used for drive transference measurement. Table 6.1 shows the parameters these measuring systems.

We can see that vacuum equipment (for example, micro lithography installations) requires the transference at nanoscale precision and it must be equipped with a laser interferometer sensor, which has the transference error $\delta_I = 10$ nm.

In case of higher precision (in the range $\delta_I \leq 0.1$ nm), which is used in scanning tunneling microscopy, the designers use measurement of tunneling current between the instrument and the sample. The tunneling variation is used to determine the distance between the instrument and the sample [3].

The transference error δ_c of the drive can be expressed as a function of static friction force according to the equation presented in Chapter 5 (see Equation (5.13)):

$$\delta_c = F_{CT}/k_0,$$

where k_0 is the drive transmission coefficient in static friction force; F_{CT} is the static resistance force in the drive (in motor, chains of transmission, in pivots).

Static resistance force of vacuum equipment is formed not only due to surface friction, but also as a result of the weight of transported objects, atmospheric pressure, and elastic forces which can be as high as hundreds of Newtons.

At the beginning of a motion, a dead space of follow-up system (error δ_c) can be determined by Equation (5.13):

$$\delta_c = I_{TP}/(k_D \cdot k_y),$$

where I_{TP} is the steering command (starting current on the drive input which corresponds to the beginning of the drive movement); k_D is the transmission coefficient of the positioning sensor; k_y is the amplifier coefficient of measuring system.

The theory of follow-up system [5] shows that dead space δ_c of any type of the drive is determined by the starting current (I_{TP}) of the follow-up system and by the common transmission coefficient of the measuring system:

$$k = k_D \cdot k_y.$$

In case of a constant motor control current the higher amplifier coefficient corresponds to a lower positioning error, but in this case the stability of the control system can be changed.

For determining the optimal value of the coefficient $k = k_D \cdot k_y$ it is necessary to create the dynamic model of the drive control [5] which should be stable, of high quality, and very precise.

The analysis of the total positioning error components of mechanisms of a precise vacuum drive shows that the most influence on the transference error is in the static error δ_{CT} of the drive, also as in the error δ_c as a result of static friction force in the drive.

As a result of different types of the drive analysis, of the matrix analysis, of the matrixes of the drive structure and of the drive design, the drive composition and drive structure variants were classified according to two main parameters: precision of transference and quick-action of the drive.

These parameters are: "static friction force" which determines error δ_c and which acts in the motor (see Figure 6.1), pilots (see Figure 6.2) and in vacuum sealing elements (see Figure 6.3).

The "type of the drive" parameter determines quick-action of the drive, see Figures 6.1 and 6.3.

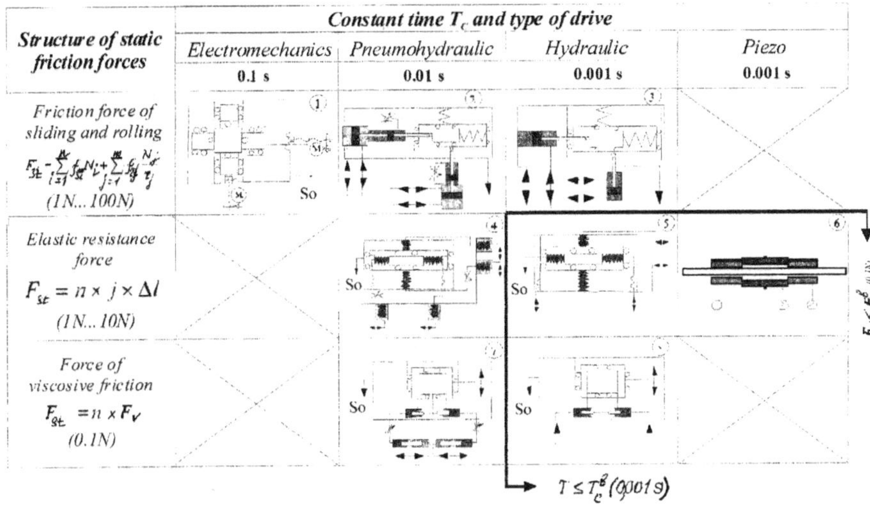

Fig. 6.1 Analysis of the precise vacuum mechanisms by parameters: "type of drive" and "static friction force in the drive".

Apart from this, the drive pilots are analyzed by the parameter "Number of movement degrees" as it is shown in Figure 6.3. In technological and in research equipment it is necessary to move the object in three or more coordinates.

Let us analyze different kinematic schemes of the drives which can be used in vacuum technological equipment, in research vacuum equipment or in other equipment which is estimated by clean technological media where the drive only meets minimal friction forces and can ensure maximal rigidity of the elements of the kinematic chain.

As is known from the theory of automatic control [6,7], hydraulic, pneumatic and pneumohydraulic drives have the best dynamic parameters. Nevertheless, historically in the beginning of vacuum equipment development, the first vacuum equipment was equipped with an electromechanical drive which had base elements operating in vacuum.

The electromechanical drives are usually supplied with long multi-element kinematic chains. These chains consist of motor, sockets, reduction gears, transmission, sealing elements (vacuum motion feedthroughs) [6]. The presence of long kinematic chains leads to decrease of its rigidity and decrease in kinematic and dynamic precision [6]. Electric motors of such drives cannot be used inside vacuum chambers of such installations without additional sealing elements, because of high level of electromagnetic fields, bad vacuum sealing and high level of the outgassing flow. The additional sealing elements and electric motor protection leads to an additional length of the kinematic chain.

Let us analyze different kinematic schemes of the drives (see Figure 6.1) using the parameter "structure of static friction force in the drive".

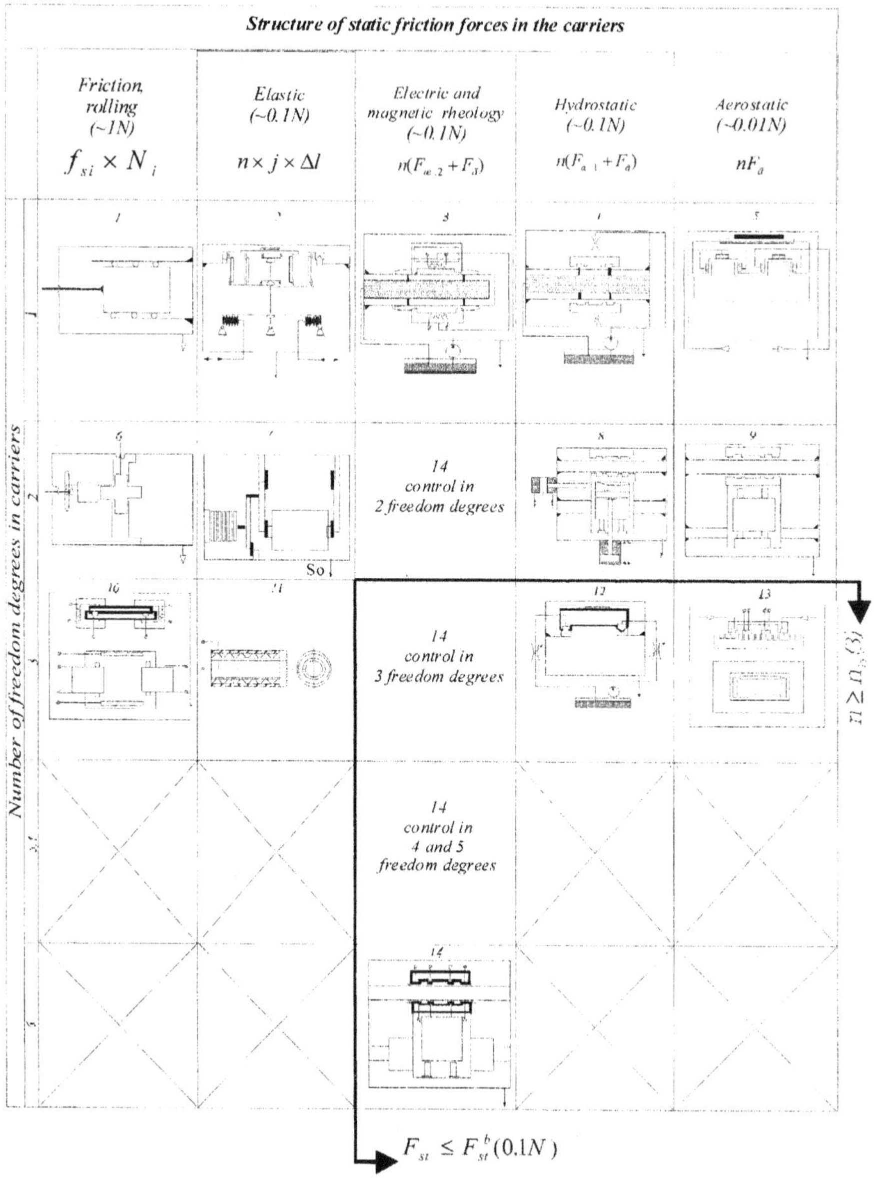

Fig. 6.2 The precision vacuum mechanisms analysis using the parameters: "static friction forces in the carriers" and "number of freedom degrees in the carriers".

Scheme 1 in Figure 6.1 shows the electromechanical drive of an electron beam lithography installation in which the drive is based on a stepper motor. The motion of the stepper motor goes through gear pairs, nut-screw pairs to low and to high

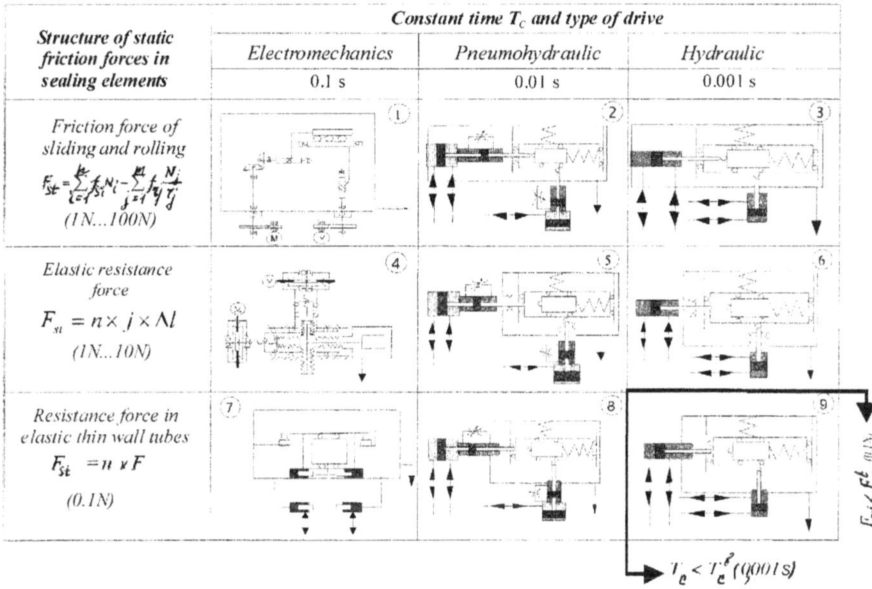

Fig. 6.3 Analysis of the precise vacuum mechanisms by parameters: "constant time T_c and type of drive" and "static friction forces in sealing elements".

pilots of coordinate table in vacuum. Thin-wall elastic sockets are used as sealing elements.

The schemes 2 and 3 show the pneumohydraulic drive and hydraulic drives based on "cylinder-piston" pairs and based on stuffing box seal type. Static friction forces appear in the seals of hydrocylinders. Furthermore, these schemes show the pneumohydraulic drive and hydraulic drives based on welded bellows.

The pneumohydraulic drive (scheme 4) consists of four welded bellows connected by pairs of tubes which are filled in by work liquid. In every pair of bellows one bellow is situated inside pneumatic germetic volume of cylinder. Resistance force in the drives 4, 5 (schemes 4, 5) appears because of the deformation of welded bellows and this force varies in value and in direction in the process of coordinate table movement. Scheme 6 shows a stepper piezo drive, which contains three piezo tubes. Resistance force of the drive appears in the middle piezo-tube in the process of its axial deformation.

Schemes 7, 8 show the variants of pneumohydraulic (7) and hydraulic (8) drives, which use thin-wall rubber socks.

In case of precise transference, the high requirements are not only on the motor, but also on pilots which as a transference base, have a great influence on the precision of the transference. The pilots of a high precision drive must comply with the following requirements:

- low friction;
- high rigidity;

- absence of recoil;
- high smoothness of the transference;
- high wear protection;
- compatibility with vacuum and other clean conditions;
- technology protection.

The design and creation of pilots which meet all of these requirements is a difficult task. In the realization of this task the designer often creates the best technical variants but these variants often compete with other requirements, and for example the designer has to use non-recoiling pilots with small friction force but with low rigidity.

In the mechanisms of high precision all types of pilots can be used as shown in Figure 6.2: friction, rolling, elastic, hydrostatic, aerostatic.

The most often used variants are the rolling (scheme 1) and the friction one (scheme 6 corresponds to electron lithography installation), which corresponds to the sample transference in one, two, or three direction [8]. The rolling type pilots correspond to small friction forces, small differences between static and dynamic friction force, high rigidity and high precision in all directions of the transference. Nevertheless, this type of pilots has high cost (price) and increased in its manufacturing, it has also small damping ability in the direction of the motion. Movable friction pilots were not used in vacuum technology till recent time. New materials and new technology can help designers of electronic technical equipment to decrease the manufacturing costs of this type of pilots and to increase the damping ability.

The success of manufacturing of friction-type pilot depends on ceramic materials, new technology, and new antifriction coatings. In engineering, designers use pilots supplied with combined facets of rolling and friction transference. Rolling facets are situated in the direction of maximum load. The other ones are covered with antifriction coatings or use Teflon coatings combined with hardened steel. This variant increases the rigidity of this type of pilot from 1.5 to 3 times in comparison with rolling type pilots. This type of pilot has lower resistance forces 4 to 7 times in comparison with friction type pilots and better damping properties. Scheme 2 shows the elastic pilot with one degree of freedom for electron beam lithography installation. The schemes 7 and 11 show piezo electric mechanisms of elastic pilots, the mechanisms of high precision transference in two and in three directions. The use of hydrostatic and aerostatic pilots is limited because of bad properties which cannot be used in vacuum and clean technological conditions. Good properties of this pilot (no recoil, small friction forces, smooth movement, high precision) should be accompanied by good sealing properties of these pilots. Scheme 4 shows the hydrostatic pilot (of cylinder form), which can be used in coordinate table drive for electron beam micro lithography installations. Scheme 8 shows two coordinate table for electron beam lithography which is supplied with hydrostatic pilots of cylinder shape. Scheme 12 shows hydrostatic pilot of one coordinate measuring system [9], which is supplied with electrohydraulic transformers which allow the realization of microtransference in vertical direction and in rotation around a horizontal axis.

These microtransferences are used for precise positioning. Scheme 3 shows hydrostatic pilot design based on magnetic rheology phenomena [10, 11]. In this design the magnetic rheology suspension is used as a work liquid. This pilot has all advantages of hydrostatic pilots:

- low friction force,
- high precision,
- high smoothness of transference,
- good damping properties,
- high rigidity,
- absence of recoil.

Active throttling of a magnetic rheology liquid helps to realize radial micro transference of a pilot body within limits of radial gaps of the pilot. The pilot sealing is realized using thin-wall rubber socks. Scheme 14 shows a two-coordinate table equipped with electro-rheology throttling of work liquid [10]. As a work liquid of this pilot an electro-rheology suspension is used.

The number of motion freedom degrees of this pilot is determined by a control system and this number can range from one to six.

Schemes 5 and 9 show one- and two-coordinate tables based on aerostatic pilots for installations of photolithography. Scheme 13 shows the coordinate table pilot which is used in photolithography installation [12] and which is based on the linear stepper motor. The table is transferred in two coordinates and is based on an air cover which is fixed on a flat plate which can be rotated on a plate surface.

A vacuum drive must produce a minimum outgassing rate of gases in a vacuum chamber or in a chamber with clean technological media. Different types of sealing elements with different structure of resistance force are used for this purpose. Let us analyze the precise vacuum drive on parameter: "Resistance forces caused by sealing elements". Scheme 1 (see Figure 6.3) shows the electromechanical drive based on throttle motors. Vacuum rotary-motion feedthrough is based on two vacuum rubber cuffs. The rotation motion is transformed into linear motion using the "shaft–roll" friction pair. Scheme 2 shows the variant of pneumo-hydraulic drive of electron beam lithography installation. The pistons and rods of pneumo hydraulic cylinders are sealed in this installation by cuff seals. Scheme 3 shows the drive in which cylinder rods are sealed by cuff seals. These drive schemes are determined by the work liquid evaporation into vacuum chamber and this contaminates the elements of electron optical system. Scheme 4 shows the electro mechanical drive of electron beam lithography installation. The drive sealing is realized by metal bellows which are protected from atmosphere pressure by T-form element. Schemes 5 and 6 show the variants of pneumo-hydraulic (Scheme 5) and hydraulic drive (Scheme 6) in which welded bellows protect the system from the vapors of work liquid. Scheme 7 shows the electromechanical drive based on four hydro-cylinders sealed by thin-wall rubber socks. The pairs of hydro-cylinders are connected by tubes. Schemes 8, 9 show the variants of pneumo-hydraulic (Scheme 8) and hydraulic drive (Scheme 9) in which the hydro-cylinders are sealed by thin-wall rubber socks.

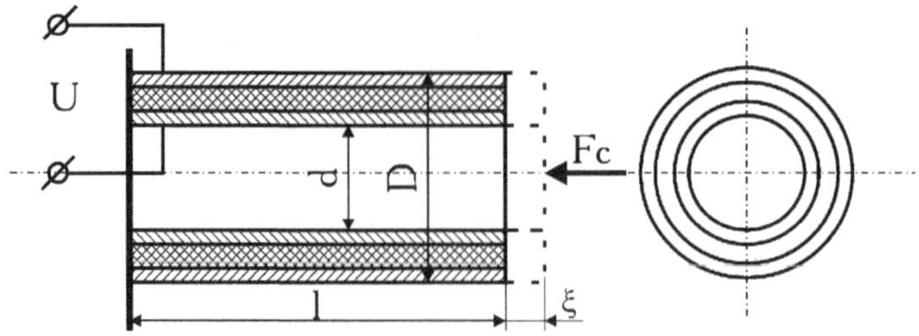

Fig. 6.4 Scheme of piezo tube drive.

As we can see, the analysis of different types of pilot and different types of sealing elements forms the basis for selection of optimal designs and structures of precise vacuum drives (Figures 6.1, 6.2, 6.3). The zones of optimal structural solutions of the drive are shown in the tables by arrows. The arrows point to some technical parameters: time constant T_Π, (the pointed drive parameter in which T_Π is less than 1 millisecond), static resistance force F_{st} (the pointed drive parameter in which F_{st} is less than 1 N). As we can see points show the zones of tables where the drives of high positioning precision are situated.

6.2 Physical Effects Which Are Used for Vacuum Mechanisms of Nanometer Precision Creation

6.2.1 Piezo Effect

Analysis of literature and patents [13–15] allows one to find piezo transformers which can be used for vacuum mechanisms of nanometer precision.

These mechanisms can be characterized by the properties:

- error of transference below ±10 nm (nanometers),
- high smoothness of transference,
- high sensitivity,
- high speed action (time constant below 2 microseconds),
- possibility of reversed transference,
- absence of heat generation at work.

The range of piezo drive transference is less than 400 mic. Piezo transformers can be used to perform two types of motion: transversal and longitudinal. Let us consider two corresponding schemes of piezo transformers: tube and column.

For the tube scheme, the transference of the end of the drive shown in Figure 6.4 can be expressed [15]:

$$\xi = 2d_{31}lU/(D-d) - 4S_{11}lF_c/(\pi(D^2-d^2)). \tag{6.3}$$

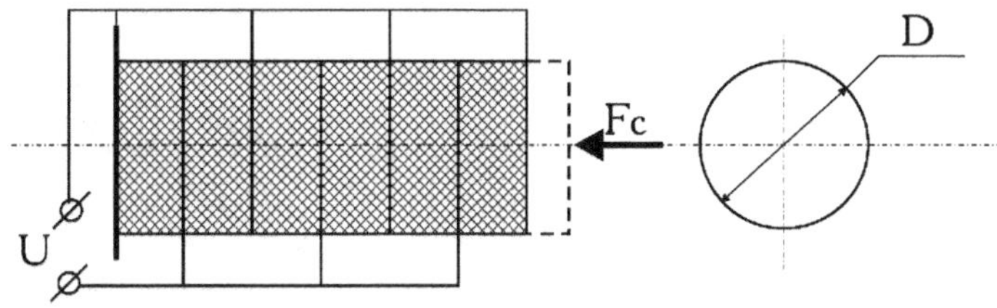

Fig. 6.5 Scheme of piezo column drive.

For the column scheme, the transference of the drive shown in Figure 6.5 can be expressed as [15]:

$$\xi = nd_{33}U - 4nS_{33}hF_c/(\pi D^2), \quad (6.4)$$

where d_{31}, d_{33} are the parameters of piezo modules corresponding to effects of cross and longitudinal deformation as a result a unit electric field m/v; l, h, d, D are the size parameters of tube and column, m; n is the number of disks in the piezo column; U is the input voltage of piezo drive, V; S_{11}, S_{33} are the backward modules of elasticity of piezo materials of tube and column; F_c is the external load, N.

The size of longitudinal piezo module d_{33} is two-fold larger than d_{31} of cross piezo module [15].

The maximum force of piezo tube $F_{c\,\max}$ (Figure 6.4) [15] can be expressed as

$$F_{c\,\max} = \pi d_{31}(D + d)U/(2(S_{11})^E). \quad (6.5)$$

The maximum force of piezo column $F_{c\,\max}$ (Figure 6.5) can be expressed as

$$F_{c\,\max} = \pi d_{33}D^2U/(4(S_{33})^E h). \quad (6.6)$$

The designers usually use piezo ceramics for piezo drives. Usually this ceramic is of the ZTP type which consists of Zr, Ti, Pb. These components have the highest piezo activity, high time stability of properties, high stability of electric load. The graph of the drive transference $X = f(U)$ (Figure 6.6) shows hysteresis which is typical for the piezo electric drive. As an example of a piezo drive let us consider a drive based on a piezo column supplied with a reduction mechanism [15]. The scheme of this drive and its main parameters (transference X of output part of the drive, piezo column deformation ξ) are shown in Figures 6.6 and in 6.7.

Figure 6.7 shows a piezo column of active length $l_A = 50$ mm which is made of ZTP piezo ceramic. It consists of discs of diameter $D = 18$ mm and thickness $h = 1$ mm. The length of the column transference is $X = 200$ mic. The voltage of input current is $U = 600$ V.

Piezo drives parameters can be described by the "after effect" which can be described by deformation of piezo element increasing at constant electric load. For example, the drive shown in Figure 6.7 reaches deformation of piezo drive about 1 ... 2.5 mic after applying constant voltage U during a time period of about 4

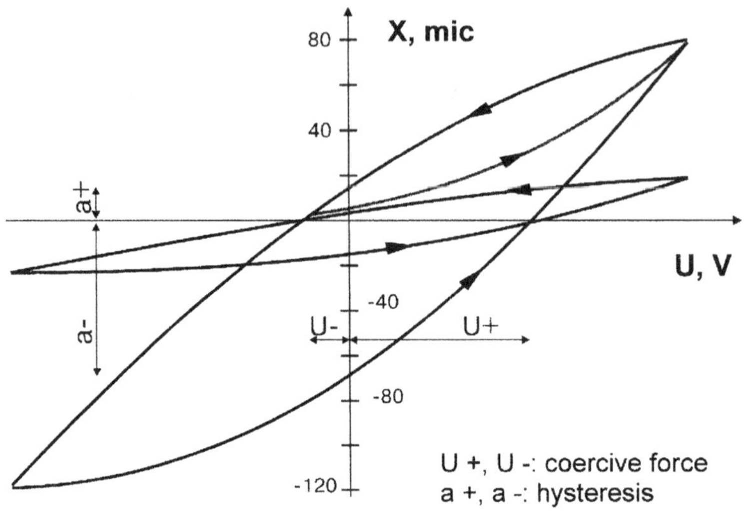

Fig. 6.6 The piezo column deformation (output element transference) X (mic) as a function of input voltage U. The symbols on the graph U^+, U^-: coercive force, a^+, a^-: hysteresis.

minutes, as shown in Figure 6.8. This effect can be removed by increasing voltage on the piezo drive.

6.2.2 Magnetic and Electric Rheology Effects

Magnetic and electric fields in cross-section of a tube (throttle) (see Figure6.9) filled with magnetic (MR) or electric (ER) rheology suspension lead to quick variations of dynamic viscosity of these suspensions (see Figure 6.10). These effects can be used for design of magnetic and electric rheology throttle (MR and ER throttle). This design helps to remove traditional for hydraulic design controlling elements: valves, slide membranes, screens. These factors allow one to decrease the service life of a throttle. The time period is less than 1 millisecond for a magnetic rheology throttle and less than 0.1 millisecond for an electric rheology throttle because of the elimination of inertia elements.

As a work liquid in these throttles magnetic rheology (MR) and electric rheology (ER) suspensions can be used. These suspensions can be considered as a dispersed physical-mechanical system consisting of two (liquid and solid) phases separated by a large surface. One of the phases (dispersed phase) is distributed in size of the particles from 10^{-3} to 10^{-2} mm.

As a dispersed magnetic rheology phase, the designers use microparticles of Fe and CrO_2 or microparticles made of other soft magnetic materials. As a dispersed

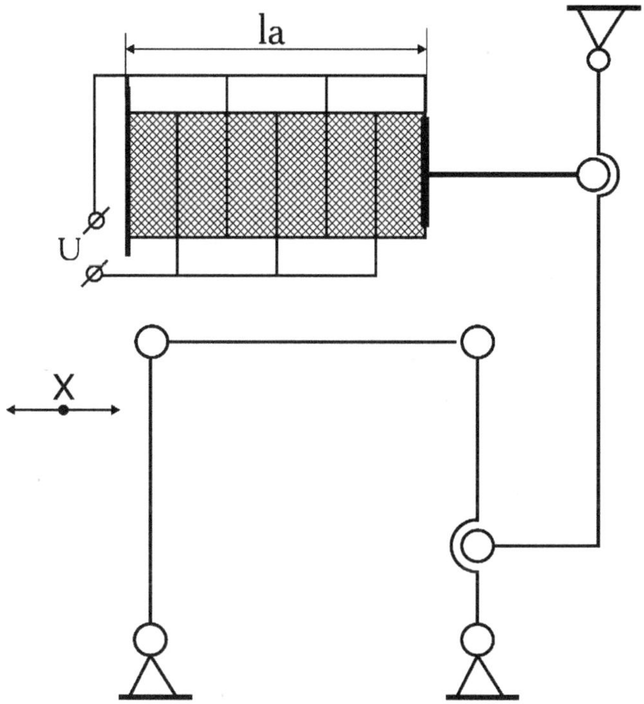

Fig. 6.7 Scheme of piezo drive supplied with reduction mechanism.

electric rheology suspension microparticles of SiO_2 can be used. As a dispersed phase vacuum oil can be used.

The stream of magnetic rheology (MR) suspension can be described by Newton's flow laws [16] with Newton behaviour parameter n. In absence of a magnetic field in the zone of low suspension speed, the flow of magnetic rheology suspension corresponds to pseudo-plastic law:

$$\tau = k \cdot \dot{\gamma}^n, \tag{6.7}$$

where τ is the shearing stress; k is the consistence of magnetic rheology (MR) suspension; γ is the shearing velocity; n is the parameter of magnetic rheology suspension in Newton behaviour.

Under external magnetic field, a magnetic rheology suspension changes its viscosity: thre are plastic properties [16, 17] in the area of low shearing velocities with behaviour corresponding to the Balkly–Gershel law:

$$\tau = \tau_0 + k \cdot \dot{\gamma}^n, \tag{6.8}$$

where τ_0 is the viscous flow limit which depends on external magnetic field, material, volume concentration of dispersed phase and magnetic rheology suspension

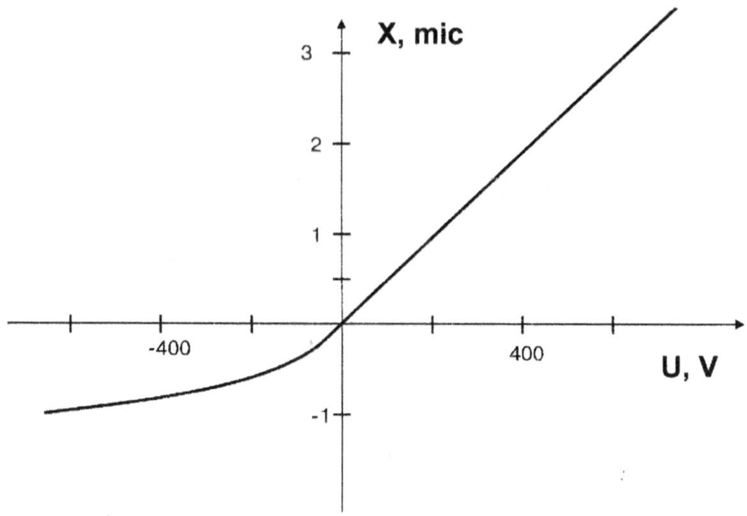

Fig. 6.8 Diagram shows "after effect" of piezo drive; the symbols on the graph: X – column transference (mic), U – constant voltage (V).

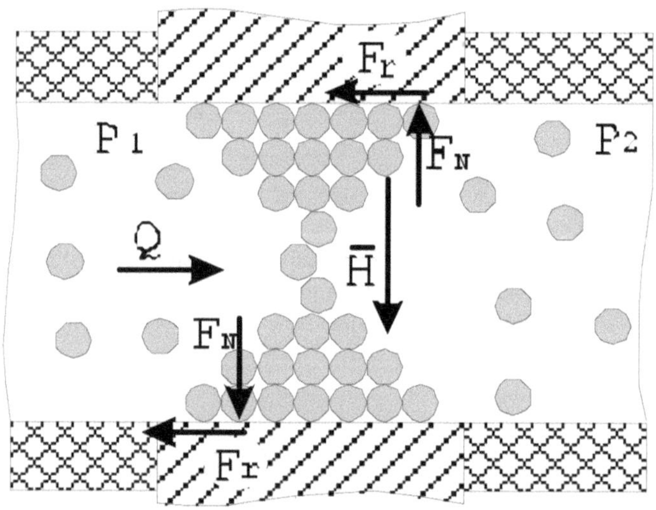

Fig. 6.9 Scheme of work gap of magnetic rheology throttle.

temperature; k is the consistence of magnetic rheology (MR) suspension, which is similar to plastic viscosity of work liquid.

We can see from Equation (6.8) that shearing stress τ depends on shearing velocity. This dependence corresponds to non-linear law. To simplify the engineering calculations of a precise hydro-drive based on magnetic rheology suspension as well as to use the automatic control methods of a hydrodrive, it is necessary to consider a magnetic rheology suspension as a Newton liquid described by Newton behaviour, which is equal to one. It is necessary also to use the term "equivalent dynamic viscosity". According to Ichibara [8], the term "equivalent dynamic viscosity" corresponds to viscosity of Newton liquid whose flow is equal to abnormal viscous liquid.

Equation (6.8) of shearing velocity $\gamma \neq 0$ in this case can be presented as

$$\tau = \mu_{\text{eqv}}\dot{\gamma}, \tag{6.9}$$

where μ_{eqv} is the "equivalent dynamic viscosity" of magnetic rheology suspension.

It is known that Newton liquid flow is described by the linear equation (6.9) if the liquid flow is layered. The critical value of Reynolds number (flow laminarity must be less than Reynolds number) usually corresponds to [18]

$$\text{Re} = u \cdot h/\nu = 1000 \ldots 1200, \tag{6.10}$$

where h is the work gap for magnetic rheology throttle; u, ν is the flow speed and kinematic viscosity coefficient of work liquid (magnetic rheology suspension). The Reynolds number is less than 10 in magnetic rheology hydrodrive. This factor ensures layered flow of work liquid.

In our case "equivalent dynamic viscosity" μ_{eqv} of magnetic rheology suspension or electric rheology suspension can be considered as a function of the following parameters: magnetic field strength H or electric field strength E, temperature t (°C), volume concentration of dispersed phase of micro particles φ_v and dynamic viscosity μ of dispersed phase. The graphs of "equivalent dynamic viscosity" μ_{eqv} of magnetic rheology suspension based on carbide Fe particles and graphs of "equivalent dynamic viscosity" of electro rheology suspension based on SiO_2 particles as a function of magnetic field strength H or electric field strength E [8] are shown in Figure 6.10.

The dispersion medium in both examples is vacuum oil VM-4 (Russian standard). The temperature of suspension is constant in both cases and is equal to $t = 30$°C. The volume concentration of micro-particles of dispersion phase is

$$\varphi_v = V_P/(V_P + V_M),$$

where V_P, V_M are the volumes of microparticles (V_P) and of oil (V_M) which are collected for magnetic rheology suspension at field strength: $\varphi_v = 0.2, 0.16, 0.12$ and for electric rheology suspension at field strength $\varphi_v = 0.2$.

Graphs of equivalent dynamic viscosity $\mu_{\text{eqv}}(H, E)$ (Figure 6.10) can be divided into three parts. The first two parts of the graphs can be described by the equations: $\mu_{\text{eqv}}(H, E) = A_0 + A_1 \cdot H(E)$, where A_0, A_1 are the constant coefficients. It can be used for the drive control because of good linear properties of the parameters. The third part of the graph is described by the large variation of $\mu_{\text{eqv}}(H, E)$ and after magnetic field strength H or electric field strength E reach the strengths values H_S or E_S (Figures 6.10b, a) the rigid "cluster membrane" is created at the gap of the

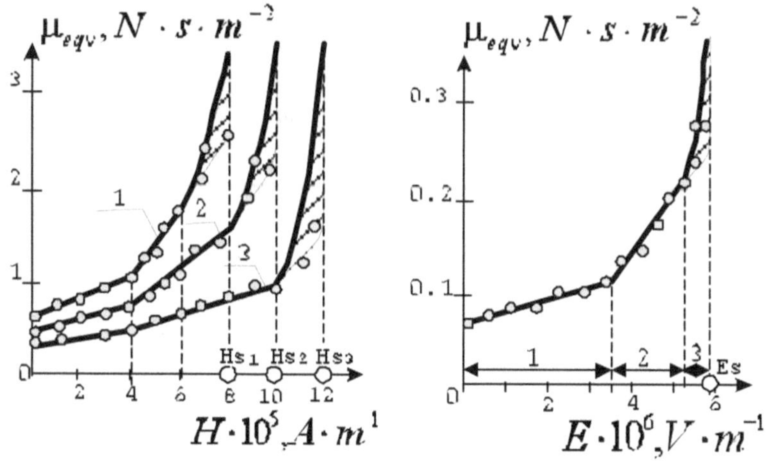

Fig. 6.10 Graphs of equivalent dynamic viscosity μ_{eqv} as a function of field strength in work gap; (a) for magnetic rheology suspension (1: $\varphi_v = 0.2$; 2: $\varphi_v = 0.16$; 3: $\varphi_v = 0.12$); (b) for electric rheology suspension ($\varphi_v = 0.2$).

throttle. This membrane shuts down the throttles based on magnetic rheology or electric rheology.

The expenditure of work liquid in throttles of magnetic rheology or electric rheology during the work process which corresponds to the first two parts of the parameter $\mu_{eqv}(H, E)$ variation can be expressed as

$$Q = (P_1 - P_2)S^3 b/(12\mu_{eqv}(H, E)L), \tag{6.11}$$

where P_1, P_2 are the input and output pressure of work liquid of magnetic rheology and electric rheology throttles; S is the value of work gap; b is the width of the gap in cross direction which is normal to the flow of the work liquid; L is the length of the gap in the direction of flow of the work liquid.

6.3 Vacuum Drives and Manipulators of Nanoscale Precision

Matrix analysis of vacuum mechanisms (see Section 6.1) allows one to find optimal structure schemes of the drives which can provide high precision, quick action, good vacuum parameters. Let us consider the design and the work of these mechanisms: piezo drives, magnetic and electro rheology drives and manipulators.

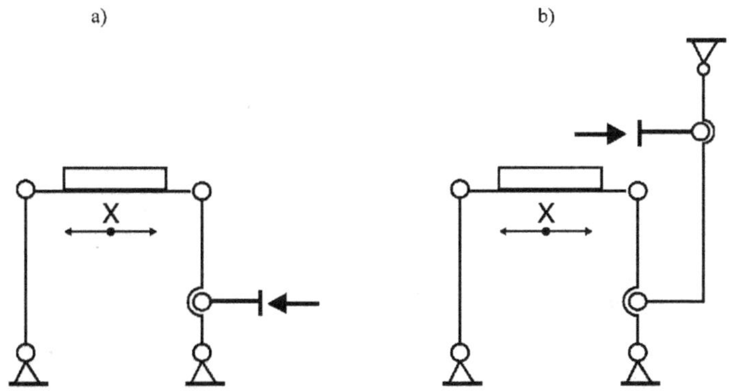

Fig. 6.11 Kinematic schemes of piezo drive of linear transference supplied with lever-multiplication system: (a) one-step system; (b) two-step system.

6.3.1 Vacuum Piezo Drives

Let us consider piezo drive based on a piezo column [15] supplied with a reductor mechanism. The scheme of the drive and its parameters: output element transference X, piezo column deformation ξ are shown in Figures 6.6, 6.7 and 6.8. Piezo column with active length $l_A = 50$ mm is made of piezo ceramic of ZTP type. The column consists of discs of diameter $D = 18$ mm and thickness $h = 1$ mm.

The piezo drive based on piezo tube [14] is used for instrument transference on three orthogonal coordinates X, Y, Z in the system of scanning microscopy.

The method presented in [15] allows one to calculate electrical parameters and design parameters of piezo drive and to providce the required output parameters of piezo drive.

The output parameters of piezo drive are the following: minimal transference $x_{\min}$ ($\varphi_{\min}$), range of transference $x_{\max}$ ($\varphi_{\max}$) in area of linear and angular coordinates. The calculated parameters are: output parameters of piezo drive ratio i of lever multiplicative system of the drive, active length of piezo element l_a, piezo element input current voltage U. The method of piezo drive parameters calculation includes the following main parts:

1. *Design of kinematic scheme of the drive*
For the required transference of output element (object): $x_{\max} \leq 80\ldots100$ mic, $\varphi_{\max} \leq 2'$ a one-step system of multiplication can be used (see Figures 6.11a, 6.12a) at $x_{\max} = 100\ldots300$ mic, $\varphi_{\max} > 2'$ a two-step system of multiplication (see Figures 6.11b, 6.12b). The ball and socket joints of lever multiplication systems (Figures 6.11a, 6.12a) must be elastic to remove gap errors.

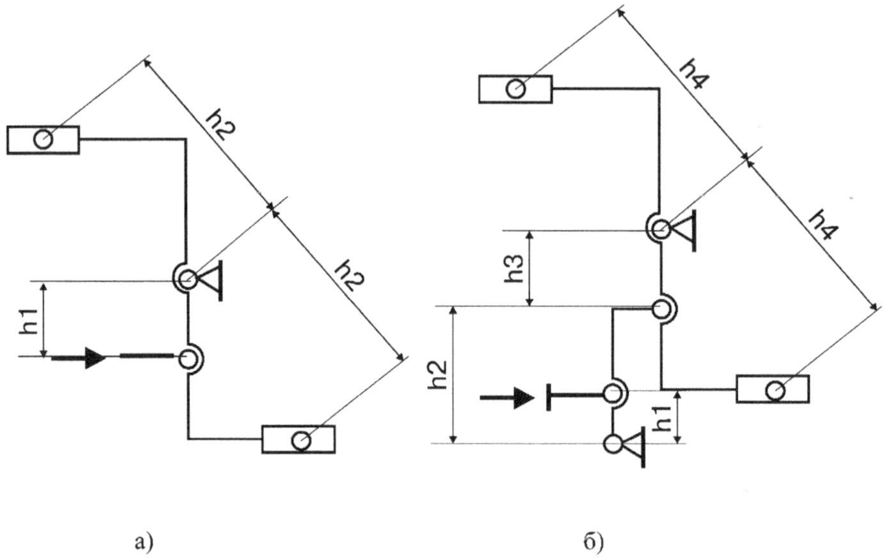

Fig. 6.12 Kinematic schemes of piezo drive of angular transference supplied with lever-multiplication system: (a) one-step system, (b) two-step system.

2. *Determination of multiplication coefficient A and axial rigidity* j_Π *of piezo element*

For piezo element in the form of piezo column consisting of discs of diameter and height $d = 18$ mm, $L = 1$ mm and made of ZTP piezo ceramic type, the parameters A and j_Π can be expressed as

$$A = 5.4 \cdot 10^{-4} \cdot l_A, \text{ mic/v}, \tag{6.12}$$

$$j_\Pi = 100.4/l_a, \text{ N/mic}, \tag{6.13}$$

where l_a is the active length of piezo element, mm ($l_a = n \cdot h$).

3. *Determination of design parameters of piezo drive*

The multiplication coefficient of a piezo drive is calculated on the base of a given transference range x_{max} (φ_{max}) in linear and angular coordinates. The equation of multiplication coefficient can be expressed as

$$\lambda(x) = x_{max}/U_{max}, \text{ mic/V}, \tag{6.14}$$

$$\lambda(\varphi) = \varphi_{max}/U_{max}, \text{ mic/V}, \tag{6.15}$$

where U_{max} is the maximum input voltage, V.

In the case when the maximum input voltage U_{max} is not fixed it can be adopted in the range of $U_{max} = \pm 600 \ldots 700$ V.

Then the multiplication coefficient A can be determined as an active length l_a of a piezo element (Figure 6.11b) and a movement translation coefficient (ratio) i.

This coefficient for one-step multiplication system (Figures 6.11a and 6.12a) is $i = h_2/h_1$ and for two-step multiplication system (Figures 6.11b and 6.12b) it is $i = h_2 \cdot h_4/h_1 \cdot h_3$.

4. *Determination of input voltage parameters of piezo drive*
The parameters of constant current of input source are the maximum output voltage $U_{\max}$, discrete of output voltage variation ΔU, output source resistance R_{out}.

$U_{\max}$ parameter determines the transference range of piezo drive. This parameter can be taken from technical conditions or obtained from the drive parameters calculation. The discontinuity of output voltage $U_{\max}$ determines the minimum value of transference.

The discontinuity of output voltage variation $\Delta U_{\min}$ determines the minimum value of transference $x_{\min}$ and can be found from the following equations:

$$(\Delta U_{\min})_x = x_{\min}/\lambda(x), \text{ V}, \tag{6.16}$$

$$(\Delta U_{\min})_\varphi = \varphi_m in/\lambda(\varphi), \text{V}. \tag{6.17}$$

The output resistance R_{out} of the work current source has an influence on the output current voltage and it can be determined after piezo element connection by

$$R_{\text{out}} = R_\Pi \cdot \Delta x \cdot m/((2\ldots4) \cdot 100), \text{ Ohm}, \tag{6.18}$$

where R_Π is the piezo element resistance at constant current, Ohm (for example $R_\Pi > 10$ MOhm); Δx is the tensile strain of piezo element, %; m is the number of piezo elements connected to the current source.

To ensure precise transference of the objects (which can be for the long travel up to hundred millimeters) in ultrahigh vacuum it is possible to use a step drive based on piezo tubes, e.g. by Burleigh Instruments, Inc. [14]. The scheme of this drive work steps is shown in Figure 6.13.

The control signal in the initial position (scheme a) is not fed to piezo tubes and these tubes are in neutral position. After control signal feeding to piezo tube 1 (scheme b) this tube is reduced in radial direction and it locks the rod. After this the control signal feeds to the tube 2 (scheme b) and this tube increases its length in axial direction. This operation leads to rod transference of one step in left direction.

After control signal feeding to the piezo tube 3 (scheme c), this tube is retracted in radial direction and this tube locks the rod. After this operation control signals of tubes 1 and 2 are eliminated one by one and it leads to width of element 1 increasing. It leads also to back movement of element 2 into initial position (schemes d and e).

After control signal feeding to the piezo tube 1 and signal elimination from the tube 3 (schemes d and e) the work cycle is finished and the drive is ready for the next step. The general view of ultrahigh vacuum step drive on the base of piezo tubes (Burleigh Instruments, Inc.) is shown in Figure 6.14. The general view of the carriage developed on the basis of the above drive is shown in Figure 6.15.

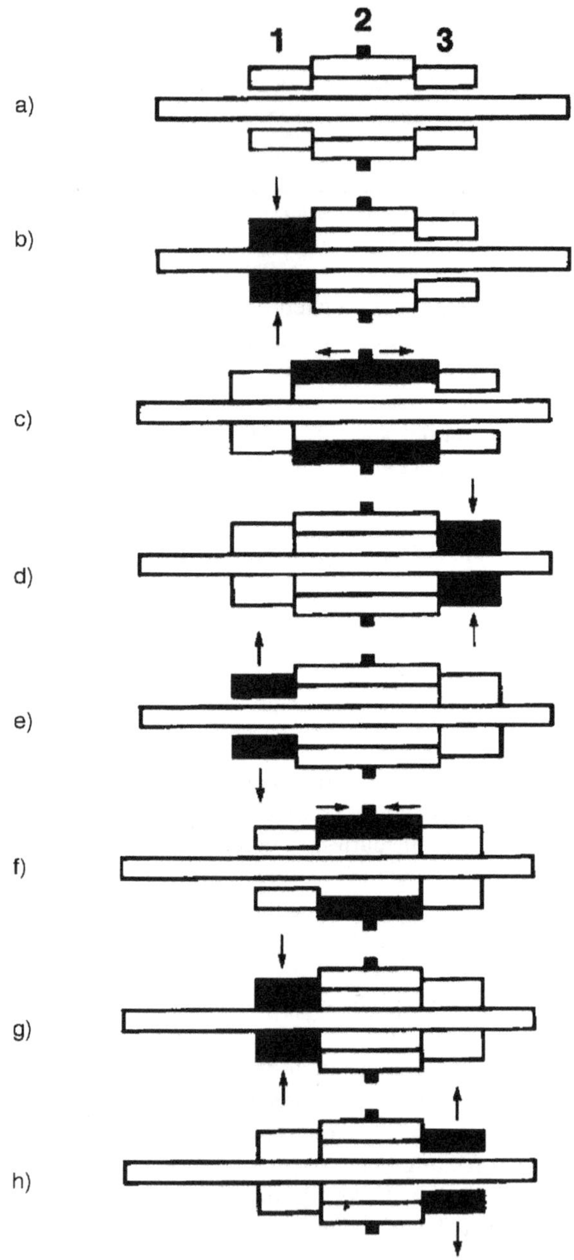

Fig. 6.13 The work stages scheme of step drive based on piezo tubes manufactured by Burleigh Instruments, Inc.

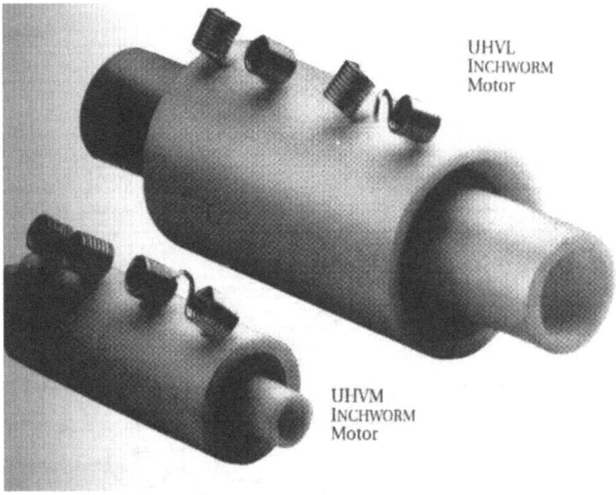

Fig. 6.14 General view of step drive based on piezo tubes manufactured by Burleigh Instruments, Inc.

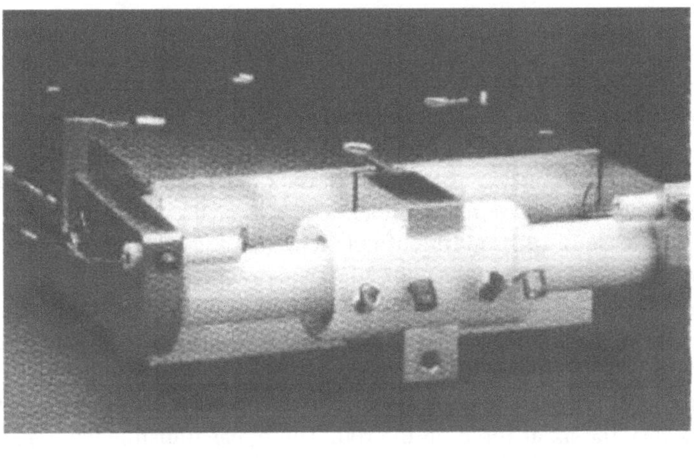

Fig. 6.15 The general view of the carriage which is done on base step drive on the basis of piezo tubes manufactured by Burleigh Instruments, Inc.

6.3.2 Multi-Coordinate Magnetic and Rheology Drives and Manipulators

Let us consider the scheme of a magnetic rheology drive with a large range of transference (up to 200 mm) in one coordinate [19] shown in Figure 6.16. The magnetic rheology suspension based on ferrum carbide and vacuum oil is used in this hydrosystem as a work liquid. The main elements of the drive are the magnetic rheological distributor 4 and thin-wall rubber seals which seal two hydro-cylinders 3

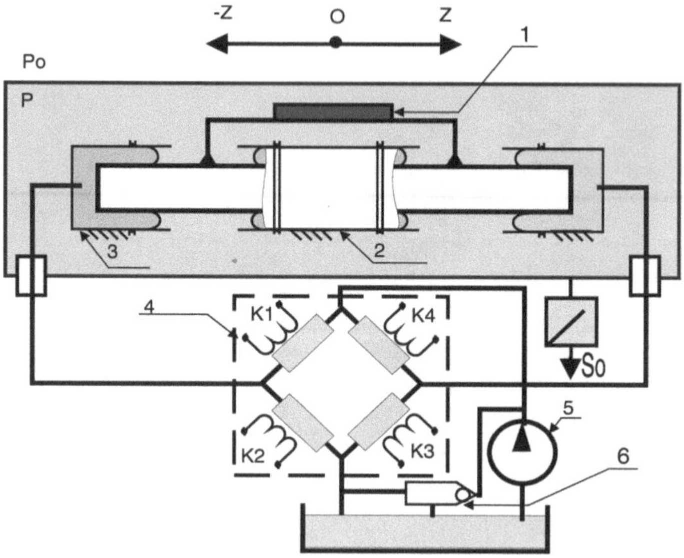

Fig. 6.16 Scheme of long travel magnetic rheology drive.

and guide 2. Magnetic rheological distributor 4 consists of four magnetic rheology throttles containing electromagnet coils $K1$–$K4$ and magnetic field conductors with work gaps for work liquid flow. The transferred object 1, guide 2 and hydro cylinders 3 are situated in a vacuum chamber. The rest of the elements are situated in atmosphere. In case of full shutting of magnetic rheology throttles, the work liquid spills over into a spillage tank through safety valve 6.

The drive works according to the following rules. The work liquid is spilled by the pump 5 from the spillage tank through magnetic rheology distributor into hydro cylinders 3. It is possible to vary the pressure in hydro-cylinders 3 in magnetic rheology throttles using control of dynamic viscosity of work liquid and, thus, it is possible to vary the axial force on the rod. The linear transference is produced in the direction of the Z axis from hydro-cylinders 3 through the rod to the carriage of coordinate table where the transferred object is fixed. In process of rod transference in the direction of the Z axis – inside thin-wall rubber seals 2; these seals roll/move from the rod to the cylinder and back. This seal ensures hermetic sealing of the connection. The general view of the drive is shown in Figure 6.17. The friction guides are used in the drive.

In the case of a small micro transference and under the following conditions: small range (up to 1 mm) of transference, millisecond quick-action, high vacuum research on technological equipment, mirror alignment of inertial mirror elements of high astronomy telescope, the precise magnetic rheology drive [14, 19] of the hydraulic type can be used. The scheme of this drive is shown in Figure 6.18. The hydrodrive consists of hydrocylinder 1, magnetic rheology throttles 2 in tubes, pump 3, sensor of transference 4 and of computer 5. The drive works in the following

Fig. 6.17 General view of long travel magnetic rheology drive.

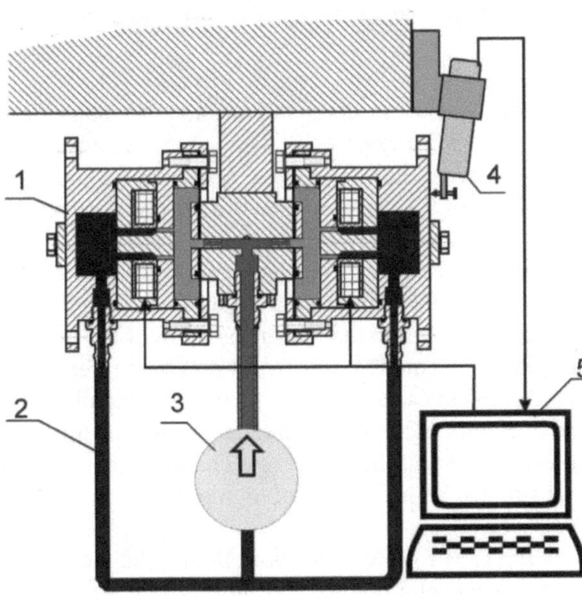

Fig. 6.18 Scheme of one coordinate tuning magnetic rheology drive.

order: work liquid-magnetic rheology suspension is spilled from the pump 3 into the hydrocylinder 1. It is possible to regulate the pressure difference of the work liquid in hydrocylinders 1 and it is possible to vary the object transference along the axis measured by the help of sensor 4.

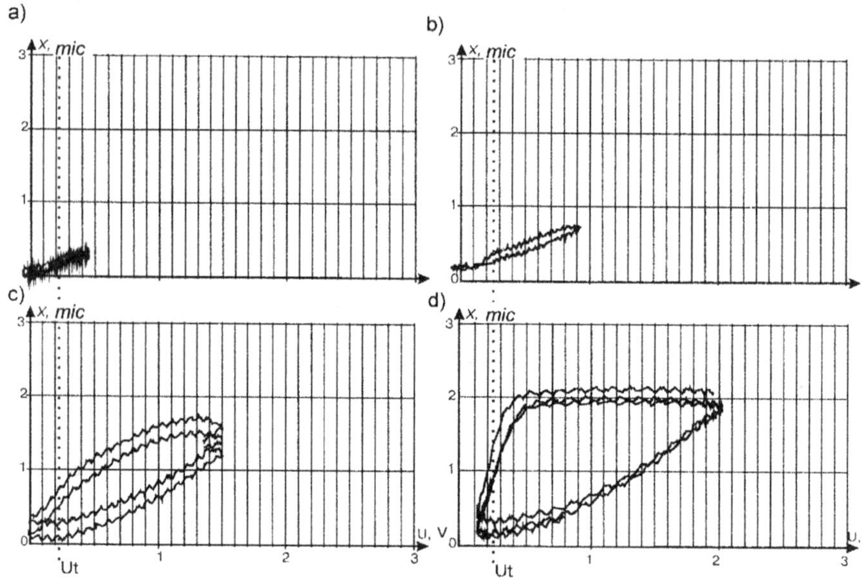

Fig. 6.19 Graphs of transference value X as a function of control voltage U.

The results of the experimental research of precision of a magnetic rheology drive (see Figure 6.18) are shown in Figure 6.19. The scheme shows graphs of transference X as a function of control current voltage U conducted to magnetic rheology throttles. The start voltage of the drive is $U_t = 0.21$ V that corresponds to the starting current $I_t = 0.014$ A. The form of the graph in Figure 6.19 shows the presence of $X = f(U)$ hysteresis (see schemes b and c in Figure 6.19). The controlling voltage in scheme c reaches its saturation zone at voltage $U = 2$ V and function $X = f(U)$. In this case it applies to the non-linear form of the graph.

For objects transference in small range (up to 1 mm) into three orthogonal coordinates X, Y, Z, it is possible to use a magnetic rheology manipulator [4], the scheme of which is shown in Figure 6.20. The manipulator contains a moving carriage 1, static base 2, rod 3 supplied with high pressure chambers, three pairs of magnetic rheology throttles 4 situated on input zone of high pressure chambers, bellow 5 and thin-wall rubber socks 6.

The manipulator works as follows. The work liquid (magnetic rheology suspension) is spilled through magnetic rheology throttles 4 into high pressure chambers of rod 3 and then it is spilled through the work gaps between the rod and the carriage to the output zone. By variation of control signal of magnetic rheology throttle it is possible to control the pressure of work liquid in chambers of high pressure and to transfer carriage 1 in three coordinates X, Y, Z. The drive is used for mirrors alignment in adaptive optical systems. The weight of the mirrors is almost 150 kg, error of transference is below 50 nanometers, quick-action (time constant) below 100 mil-

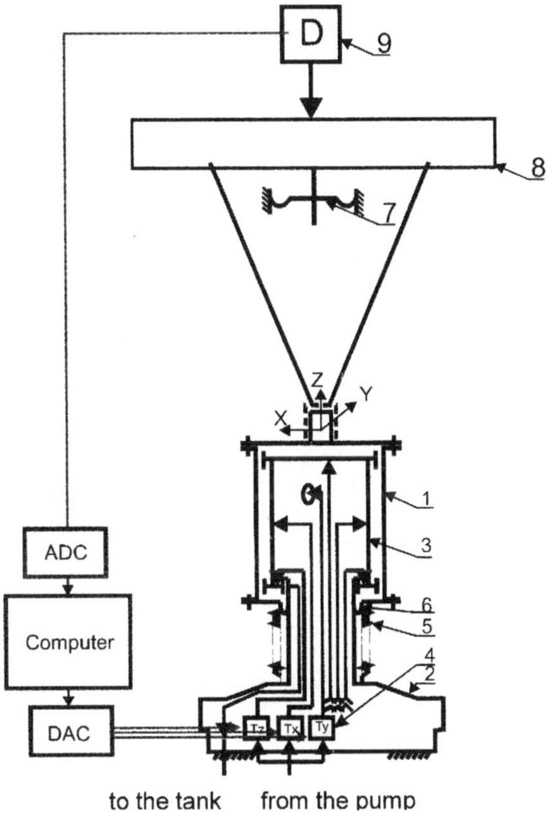

Fig. 6.20 Scheme of three coordinate tuning magnetic rheology drive.

liseconds [4]. The general view of the magnetic rheology manipulator is shown in Figure 6.21.

The equipment of microlithography, electron beam and X-ray lithography and synchrotron radiation require the object transference in three coordinates or more. Magnetic rheology manipulator (see scheme in Figure 6.22) can provide object transference in one long travel coordinate Z (travel range up to 200 mm) and in two coordinates X and Y (travel range up to 200 mic). It can also turn around axes X and Y (travel range up to 3 angular minutes) [11]

The drive contains hydro pump 1, static body 2 and moving cylinder rod. The drive is built in the form of hydrostatic way of the cylinder type supplied in a body 2 with four chambers of high pressure which form a basic element of a guide. To ensure the stability of moving cylinder rod the guide contains two basic elements fixed along axis Z. Magnetic rheology throttles in the input zone of the high pressure chambers are supplied with electromagnetic coils K1–K4, magnetic rheology throttles in the output zone are supplied with electro magnetic coils K6–K7 (coils

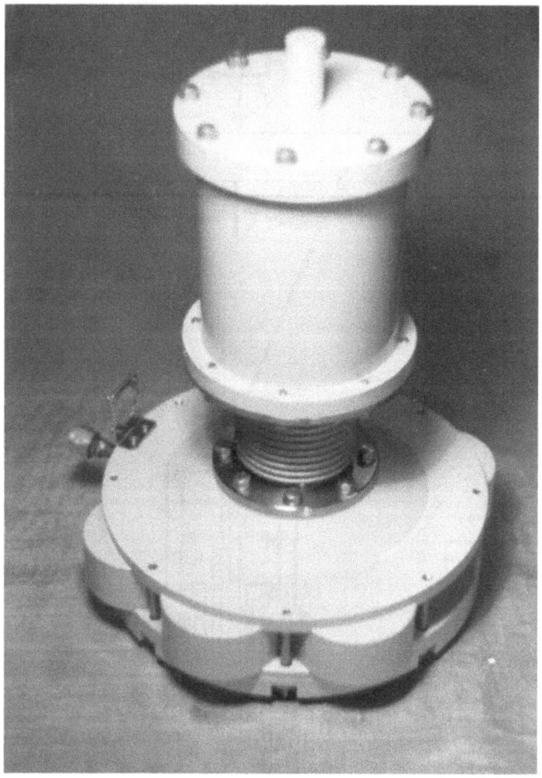

Fig. 6.21 General view of three coordinate tuning magnetic rheology drive.

K5–K7 are not shown). Magnetic rheology manipulator is sealed by thin-wall rubber socks. The drive works as follows: the work liquid (magnetic rheology suspension) is spilled from the tank by the pump 1 through magnetic rheology throttles supplied with electro-magnetic coils K1–K4. The work liquid is spilled into the high pressure chambers from the spillage tank and runs through radial work gaps between the cylinder rod and magnetic circuit of output magnetic rheology throttles supplied with coils which are supplied with coils K–5, K–8 and then it runs to the spillage tank. Using electrical current variation in coils of magnetic rheology throttles it is possible to vary pressure in the chambers of the drive. Therefore, the moving rod raises in the body 2 and the rod transfers along axes *X* and *Y* and is also transferred around axis *Z* by the long travel drive (this drive is not shown in the scheme). The general view of six-coordinate magnetic rheology drive based on hydrostatic guides and which is used for microlithography is shown in Figure 6.23.

The further idea of the use of magnetic rheology effect for vacuum mechanisms of nanoscale precision consists in the development of vacuum mechanisms based on the electric rheology effect. The electric rheology effect helps to increase the transference precision of the drive. It also helps to improve its dynamic parameters, to

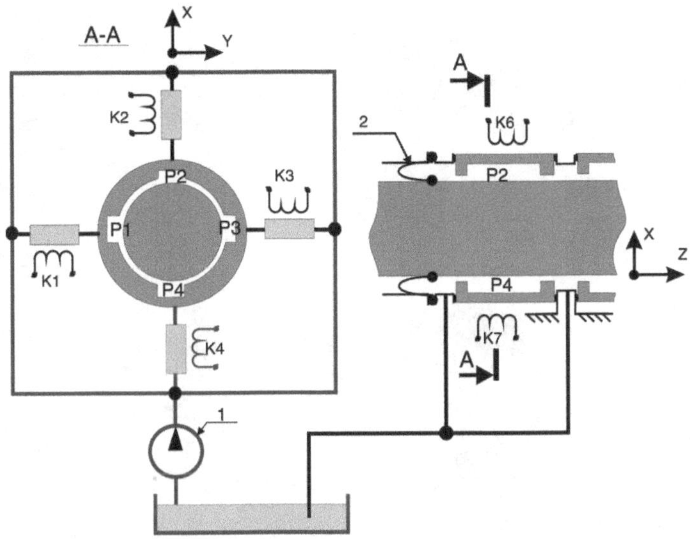

Fig. 6.22 Scheme of five coordinate magnetic rheology manipulator with long travel transference.

decrease its sizes and to decrease the weight of the vacuum mechanism. The main element of this drive is an electric rheology throttle (see Figure 6.24) which is realized in the form of a flat gap or in the form of a circular gap between metal surfaces supplied with a controlling high voltage. Variation of high voltage of electric field tension enables to vary dynamic viscosity of the work liquid (electric rheology suspension), which leaks through the gap. An electric rheology throttle allows one to exclude electromagnetic coils and to exclude magnetic circuits which are used in a magnetic rheology drive.

Let us consider a scheme of five coordinates – electric rheology manipulator [19] equipped with a loop-controlled system (see Figure 6.24). The manipulator contains input electric rheology throttles 1 and output electric rheology throttles 2, optical sensors of transference 3, control blocks 4. The manipulator is done in a form of hydrostatic guide of cylinder-shape. This guide is supplied with a static rod which contains chambers of high pressure and output electric rheology throttles 2. The mobile body with transferred object is sealed by of thin-wall rubber socks. The work liquid (an electric rheology suspension) is pumped by the pump through input electric rheology throttles 1 and through work radial gaps of output electric rheology throttles 2 and then into spillage tank. The control voltage of control block 4 is regulated on sockets of electric rheology throttles. This voltage of control block regulates pressure in the chambers of high pressure. The manipulator body rises in a layer of work liquid in the process of control. The manipulator body can be transferred in the direction of the X, Y, Z axes (in a range of up to 200 mm) and can be rotated around the axes X, Y (about 3 angular minutes). Let us estimate the components of total error of positioning $\delta\varphi_\Sigma$ (see Equation (5.10)) in case of magnetic

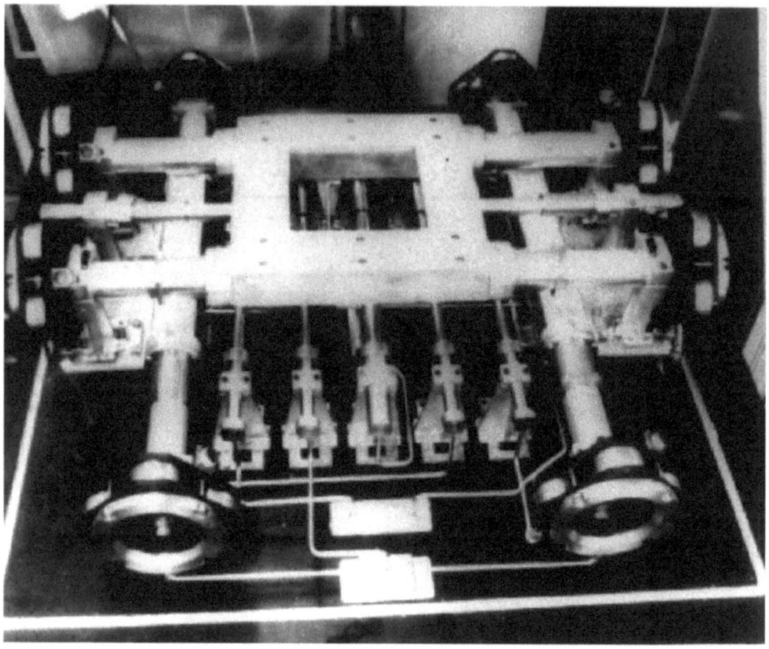

Fig. 6.23 General view of six coordinate magnetic rheology manipulator of microlithography installations.

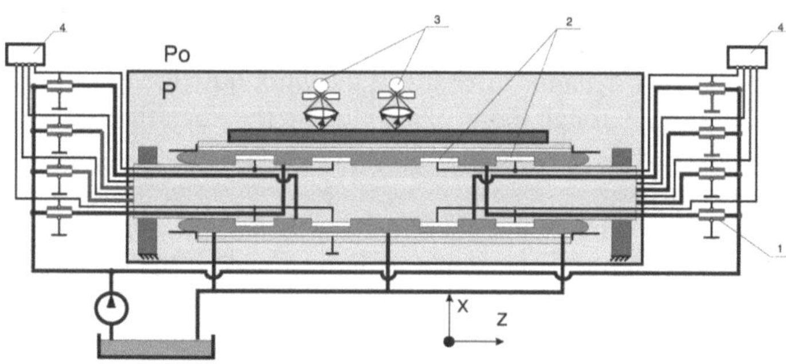

Fig. 6.24 Scheme of five coordinate electric rheology manipulator.

rheology manipulator with thin-wall socks (Figure 6.16). The transfer function of automatic loop control system of this manipulator can be estimated as

$$W_{\Pi}(S) = k_{\Pi}/(S(T_1 S + 1)(T_2 S + 1)), \tag{6.19}$$

where k_{Π} is the common gain factor (multiplication factor) of the drive; T_1, T_2 are the constant time factors of drive elements; S is the Laplace variable.

As we can see from Equation (6.19) the automatic control system of manipulator is a static automatic control system which contains one integrating chain element.

The error of driving action variation δ_V in the drive static control system (at level $\nu = 0$) is the drive error of driving action variation $\delta_V = C_0\varphi$. The drive static control system which contains one integrant element ($\nu = 1$) the positioning error is equal to zero ($C_0 = 0$).

Thus, as we can see from Equation (6.1), the positioning error is equal to zero and the error of driving action variation component depends on the speed dynamic error:

$$\delta_B = C_1\omega,$$

where ω is the stabilized speed of the drive linear transference along axis Z; C_1 is the speed error coefficient. This coefficient can be determined as

$$C_1 = 1/k, \tag{6.20}$$

where k is the common speed transmission coefficient of automatic control system of an automatically controlled drive. There is no acceleration current in this case. In case of magnetic rheology manipulator: $k = 6.282 \cdot 10^3$ 1/s and $C_1 = 1/(6.282 \cdot 10^3) = 1.59 \cdot 10^{-4}$ s.

In case of a stable transference of magnetic rheology manipulator at "small static speed", we have $\omega = 0.5 \cdot 10^{-4}$ m/s and the "component of speed dynamic error" is $\delta_V = 7.95 \cdot 10^{-9}$ m.

In case of disturbing forces (F_{dist}) (technology forces variation, vibration of base forces, variation in atmosphere pressure forces after the vacuum chamber is opening, energy variation and hydraulic pumping station variation) the error δ_N appears. This error depends on initial conditions.

Let us find this manipulator work error in case of manipulator work in positioning point. In this case all coils K1–K4 are supplied with maximal control current ($I =$ 1 A) and all magnetic rheology throttles are closed. Error δ_N in this case depends of tubes rigidity. The tubes situate between magnetic rheology distributor and hydro-cylinders supplied with thin-wall rubber tubes. Error δ_N depends also on the work liquid, on tubes of hydro-cylinders and on thin-wall rubber tubes rigidity:

$$\delta_N = F_{\text{dist}} \cdot C_r / S_{\text{eff}}^2, \tag{6.21}$$

where C_r is the total coefficient of tubes rigidity, thin-wall rubber elements, work liquid rigidity; S_{eff} is the effective area of thin-wall rubber elements in hydro-cylinders. So, in the case of disturbing forces (F_{dist}) (technology forces variation, vibration of base forces, atmosphere pressure forces variation after the vacuum chamber opening, energy variation and hydraulic pumping station variation) the error is equal to $\delta_N = 0.1 \cdot 1.342 \cdot 10^{-14}/(4.9 \times 10^{-4})^2 = 5.59 \cdot 10^{-9}$ m. The error caused by static friction forces in the manipulator can be expressed by Equation (5.13) and in the case of resistance force $F_{TP} = 0.05$ N it is equal to

$$\delta_c = 0.05/10^6 = 5 \cdot 10^{-8} \text{ m}.$$

Recoil error δ_L in magnetic rheology manipulator is equal to zero. In case of using a laser interferometer the instrumental error is equal to $\delta_I = 10^{-8}$ m. The total error of positioning is equal to $\delta\varphi_\Sigma = (7.95+5.59+50+10)\cdot 10^{-9} = 73.54\cdot 10^{-9}$ m and a magnetic rheology manipulator of this precision can be used for previous ("rough") transference of objects. Analysis of precision, dynamic and vacuum parameters of precise vacuum manipulators performed in Sections 6.1–6.3, shows that vacuum manipulators with the highest parameters are the vacuum manipulators based on piezo transformers and the manipulators of hydraulic type based on magnetic rheology and electric rheology effects. Piezo tubes allows one to transfer objects in ultrahigh vacuum ($p \leq 5\cdot 10^{-8}$ Pa) with error of positioning less than 0.1 nm in the range of transference of up to 100 mic.

Magnetic rheology manipulator and electric rheology manipulator can be used to transfer objects in high vacuum and in ultrahigh vacuum ($p \leq 5\cdot 10^{-8}$ Pa); in large ranges of transference of up to 200 mm, and to ensure an error of positioning less than 10 nm. They have good dynamic parameters: the time constant of a magnetic rheology drive is less than 100 milliseconds, the time constant of an electric rheology drive is less than 1 millisecond. They also have good damping parameters.

References

1. Landberg E.E., Mrrja C.L., Hewlett Packard, No. 5, 1981, pp. 16–18.
2. Wilson A.D., *Solid State Technol.* **29**(5), 1986, 249–255.
3. Eltsov K.N., Klimov A.N., Priadkin S.L., Shevlyuga V.M., Yurov V.Y., *Phys. Low-Dim. Struct.* **7/8**, 1996, 115–126.
4. Deulin E.A., Mikhailov V.P., Eliseev O.N., Sytchev V.V., *Proceedings of SPIE* **4003**, 2000, 303–310.
5. Smirnova V.I. et al., *The Base of Design and Calculation of Computer Regulated Drives*, Mashinostroenie, Moscow, 1983, 285 pp.
6. Ivashenko N.N., *Automatic Control*, Mashinostroenie, Moscow, 1978, 608 pp. [in Russian].
7. Makarov I.M., Menski B.M., *Linear Automatic Systems*, Mashinostroenie, Moscow, 1982, 504 pp. [in Russian].
8. Ichibara S., *Bull. Japan Soc. of Precision Engineering* **21**(1), 1987, 1–8.
9. Ravva J.S., *New Decision in Lathes Precision Increasing. Adoptatation of the System with Complex Friction*, Knijnoe Izdatelstvo, Kuibushev, 1974, 214 pp.
10. Deulin et al., *Vacuum* **44**(5–7), 1993, 469–470.
11. Anisimov V.V. et al., *Vacuum* **47**(10), October 1996, 1163–1165.
12. Panfilov Y.V., Ryabov V.T., Tsvetkov Y.B., *Equipment for Integral Schemes Manufacturing*, Radio and Sviaz, Moscow, 1988, 320 pp. [in Russian].
13. Chen V.K., *Journal of Physics* **11**(1), 1978, 1092–1093.
14. *The Power of Precision in Nanopositioning*, Burleigh Instruments, Inc. Burleigh Park, Fishers, NY 14453-0755, USA, 1997.
15. Yaffe B.I., *Piezo Electric Ceramic*, Mir, Moscow, 1974, 288 pp. [in Russian].
16. Shulman Z.P., Kordonski V.I., *Magnetic Rheological Effect*, Nauka and Technika, Minsk, 1982, 184 pp. [in Russian].
17. Bibik E.E., Some effects of microparticles mitual action in a process of ferrum liquid flow in magnetic field, *Magnetic Hydrodynamic* **3**, 1973, 25–32 [in Russian].
18. Bashta T.M., *Machine Building Machines*, Mashinostroenie, Moscow, 1971, 672 pp.
19. Anisimov V., Deulin E.A., Mikhailov V.P., Kasperski V., Sytchev V.V., *Proceedings of SPIE* **2871**, 1996, 641–648.

Chapter 7
Ultrahigh Vacuum Rotary-Motion Feedthroughs

The angular precision of the most widespread bellows-sealed rotary-motion feedthroughs is about 10 and in many cases that is not enough for the user. That is why many designers create more precise feedthroughs. An example of more precise gear is the airtight harmonic drive rotary-motion feedthrough.

This type of drive does not have a wide application, because the manufacturing process is technically complex. The flexible thin-wall gear (barrel) is the main base detail of the drive; that is difficult to master. Another reason is a short life span of this drive in vacuum.

The authors solved both problems and reduced labour-intensiveness of thin-wall gear, they made this labour-intensiveness equal to labour-intensiveness of hydro formed bellow manufacturing: they also found optimal design parameters of gear couple of harmonic drive which ensure its longevity more than $3 \cdot 10^6$ cycles of work.

7.1 Analysis of Design Variants of Thin-Wall Sealing Elements on Parameter "Manufacturability"

Comparative analysis of more than 15 variations of the hermetic harmonic drive rotary-motion feedthroughs shows that the main difference between the various drives consists in a difference of the jacket form of thin-wall flexible gears. The form difference of jackets reflects the attempts by the designers to optimize the form of thin-wall flexible gears in the following parameters:

- minimum gas flow Q_p of gases penetrated through the wall,
- maximum work longevity T_f of thin-wall flexible gear,
- high rotation rigidity J_τ,
- maximum radial and axial rigidity J_r,

E.A. Deulin et al., *Mechanics and Physics of Precise Vacuum Mechanisms*,
Fluid Mechanics and Its Applications 91, DOI 10.1007/978-90-481-2520-3_7,

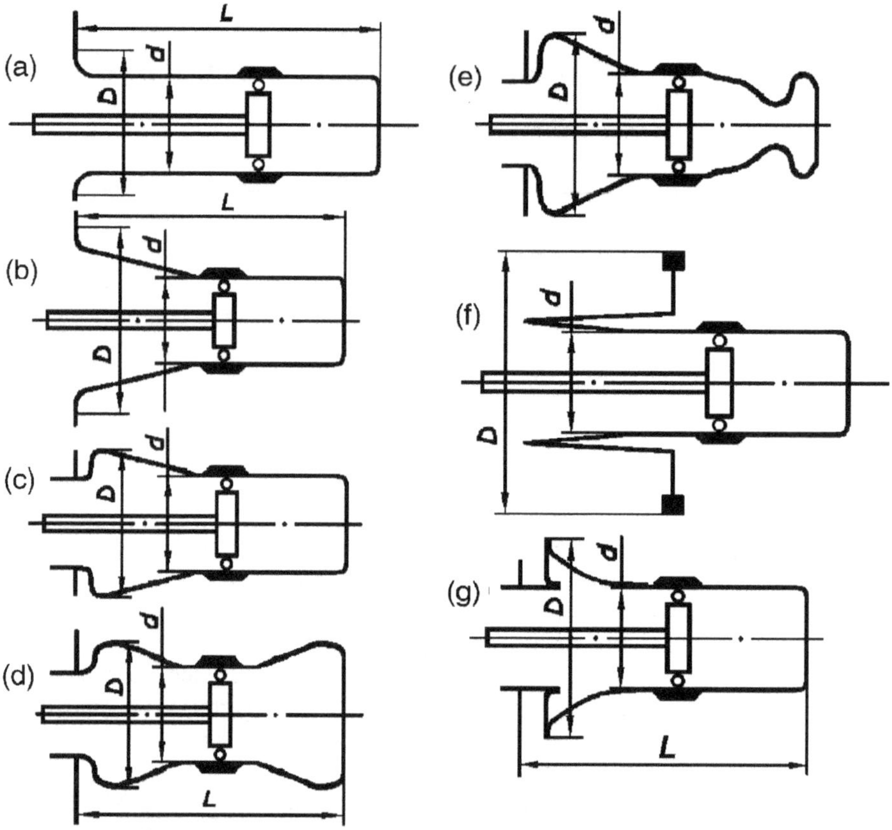

Fig. 7.1 Composition schemes of different forms of thin-wall flexible gears.

- minimum warp φ_r of thin-wall flexible gear deformation (Viennese) in a process of radial deformation,
- easy manufacturability and low labour-intensiveness C_m in a process of thin-wall flexible gear manufacturing,
- minimal overall dimensions and weight G_s.

All variations of thin-wall flexible gears may be reduced to seven basic variants (Figures 7.1a–g), the other properties are different:

Variant I (Figure 7.1a) [1]
The form of this design corresponds to the best manufacturability, this design is made in the form of thin-wall flexible gear which is designed in the form of a cylinder barrel without flanging.

The disadvantage of this design is: increased overall dimensions, which are essential for the limits to ratio L/d (where L is length of thin-wall barrel, d is mean diameter of thin-wall gear), to ensure:

- high bending stress in zone of thin-wall barrel fixation to drive body,
- absence of zigzags of thin-wall barrel form ensures high bending stress of thin-wall barrel in a deformation process;
- low longevity of thin-wall cylinder barrel as a result of its deformation by the generator.

Variant II (Figure 7.1b) [2]
The form of this design corresponds to a combination of a cylinder (right) with a cone (left) and the place of its coupling is near gear Viennese. This variant is more compact in axial direction and this structure type is more stable than variant 1. Disadvantages of this variant are:

- increased radial and overall dimensions;
- increased warp φ_r of thin-wall flexible gear Viennese, because of different rigidity of left and right parts of the thin-wall cylinder barrel as a result of its radial deformation by the generator.

Variant III (Figure 7.1c, see also Figure 2.2, Chapter 2) [2]
The form of this design corresponds to the results of matrix analysis [2]. This type of matrix analysis (developed by the authors) was described in Chapter 4, which is shown in Figure 4.7. The form of this design is different from variant I and II and the main parameter of difference is the decreased diameter of the base part of the flexible thin-wall cylinder barrel. The ratio D/d is the same as in previous variants (where D is the maximum diameter of flexible thin-wall gear).

The axial dimension of this variant is less than in variant 1 and radial dimension is less than in variant II. Variant III corresponds to an optimal geometry of the gear couple.

The main advantages of this variant are:

- small dimensions;
- high longevity of all elements of the drive;
- easy manufacturability and low labour-intensiveness in a process of thin-wall flexible gear manufacturing. The labour-intensiveness of thin-wall flexible gear manufacturing is close to the labour-intensiveness of the production of a stainless steel bellow.

The only disadvantage of this variant is the small warp φ_r of thin-wall flexible gear Viennese. This parameter is close to Variant II.

Variant IV (Figure 7.1d) [4]
The form of this design corresponds to the most rigid thin-wall of a flexible cylinder barrel. This variant ensures absence of warp φ_r of teeth of thin-wall flexible gear Viennese. This variant ensures also linear teeth transfer in the gearing process.

The disadvantage of this variant is the necessity to manufacture rigid (output) gear in the form of two sectional parts. The result of this design is, that it is difficult to produce and has a high labour-intensiveness in the process of thin-wall flexible gear manufacturing. The other disadvantage of this variant is awkwardness of design. The dimensions of this thin-wall flexible cylinder barrel (of this variant) are rather small.

Variant V (Figure 7.1e) [5]
The form of this design is done as a combination of the left part of the bell form with a special right part in a toroidal form. Both parts are attached to flexible gear Viennese. The form of the left part ensures radial rigidity, equal to rigidity of the left part of a thin-wall flexible cylinder barrel. The scheme is of small dimensions and ensures linear transference of teeth in gearing process. The scheme ensures high longevity of the flexible gear and of the thin-wall flexible cylinder barrel. The main disadvantage of this variant is it is hard and labour intensive to produce, because of a second zigzag in the form of the thin-wall flexible cylinder barrel.

Variant VI (Figure 7.1f) [6]
The form of this design is unsuccessful, because of its short life span. The reasons for this unsuccessful design are:

- tendency of its designers to ensure low cost and low labour-intensiveness of the process of thin-wall flexible gear manufacturing;
- welding junction of thin-wall barrel in a zone of high axial stress of thin-wall flexible cylinder barrel. It is a reason of small longevity of this type of feedthrough;
- increased weight and increased dimensions because of second sealing flange in the design of feedthrough;
- high labour-intensiveness of thin-wall flexible gear manufacturing.

Variant VII (Figure 7.1g) [7]
The form of this design is unsuccessful, because of its short life span. The reasons why it is so short lived are:

- two welding junctions of thin-wall barrel, one in zone of high axial tangent stress of thin-wall flexible cylinder of barrel. The other one in a zone of high bending stress of the thin wall. It is a reason of small longevity of feedthrough.
- small increased warp φ_r of thin-wall flexible gear Viennese as a result of radial deformation of thin-wall flexible cylinder of barrel,
- high dimensions and high weight of the drive because of second flange presence,
- high labour-intensiveness of thin-wall flexible gear manufacturing.

Table 7.1 shows the results of an (estimated) comparison of thin-wall flexible gear design. The estimation is done by a three point system:

1. poor quality of parameters;
2. middle quality of parameters;
3. high quality of parameters.

The analyzed parameters are:

- dimensions G_s;
- labour-intensiveness of thin-wall flexible gear manufacturing C_m;
- radial rigidity J_r;

Table 7.1 Comparative estimation of thin-wall flexible gear design.

Variant	G_s	C_m	J_r	T_g	φ_r	T_f	$\sum$
I [1]	1	3	1	2	2	1	10
II [2]	1	3	2	2	1	2	11
III [2]	3	3	3	3	2	3	17
IV [4]	1	1	3	2	3	3	13
V [5]	3	1	3	2	3	3	15
VI [6]	1	2	2	1	2	1	9
VII [7]	1	2	2	1	2	1	9

- longevity of summarized thin-wall flexible gear Viennese T_q;
- warp φ_r of thin-wall flexible gear Viennese;
- longevity of the tiredness of the drive T_f.

The left column of Table 7.1 contains the number of the different variants of thin-wall flexible gears and the number of reference which describe the analyzed variant.

Comparison of comparative summarized estimation of different variants of thin-wall flexible gear design Σ shows that parameters of drive variants which were designed for military space objects [1, 2, 4, 5] are better than the parameters of the drives which were designed for the electronics industry [6, 7]. Analysis also shows that the designers of these last variants [6, 7] try to create their variants at low costs and with little labour costs.

7.2 Precision of Harmonic Gear Rotary Feedthroughs

The authors created a theoretical model of summarized error of hermetic harmonic gear rotary feedthrough forming. The maximum kinematic error (F) of harmonic gear rotary feedthrough was researched and considered in the role of main part of summarized error and it was researched as a function of kinematic errors of main details of the drive:

$$F = \left\{k_V \left[\tan\alpha_w \cdot (E_d + F_f + F_r)\right] + k_t \cdot j_t\right\} k_m, \tag{7.1}$$

where k_V is the coefficient of kinematic error relative decreasing of gapless harmonic gear rotary feedthrough; α_w [degree] is the angle of action of harmonic gear rotary feedthrough; E_d [mic] is the error of output shaft being lead which is formed by radial beat of generator in respect to rigid; F_f [mic] is the accumulated teeth step error of flexible gear of harmonic gear rotary feedthrough; F_r [mic] is the accumulated teeth step error of rigid gear of harmonic gear rotary feedthrough; k_t is the coefficient of one teeth error transformation into error of output shaft; j_t [mic] is the recoil in single tooth contact; k_m is the coefficient of harmonic gear rotary feedthrough kinematic error relative decreasing as result of load influence.

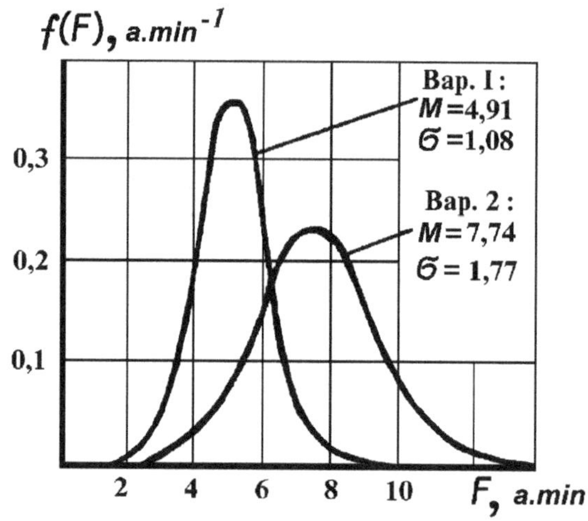

Fig. 7.2 The results of research.

It was theoretically researched and it was experimentally controlled precision quality of details of gear rotary feedthrough and relative discrimination component of kinematic errors of details influence on harmonic gear rotary feedthrough kinematic error was determined.

The results of research are shown in Figure 7.2 in form of error F frequency distribution for two sets of harmonic gear rotary feedthroughs.

The harmonic gear rotary feedthroughs investigated were manufactured in quality of precision 6–7 (graph 1) and in 7–8 quality of precision (graph 2).

The results show that the quality of precision of three basic details, changes the summarized kinematic precision of the harmonic gear rotary feedthroughs in a wide range.

7.3 Longevity of Harmonic Gear Rotary Feedthrough

The research of workability of harmonic gear rotary feedthroughs gear couples in high oilless vacuum at residual pressure less than 10^{-4} Pa, show that hard lubricant and hard coating based on MoS_2 usage increases the gearing workability in 5–20 times in comparison with its work without lubricant.

Research has shown that:

- the charging coverage of gears teeth with MoS_2 dust leads to abrasive wear of teeth;
- the coverage of gears teeth with hard lubricant based on plasma covered MoS_2 film leads to coverage wear without harmonic gear rotary feedthrough thin-wall gear teeth wear.

The longevity of tooth thin-wall flexible gear Viennese depends of the linear wear U_0 per one cycle of gearing (beginning and end of one cycle of gearing) of hard solid teeth coverage in critical point of teeth. The gearing cycle corresponds to a half of a shaft revolution (turn).

$$U_0 = 16.3 \cdot \frac{k \cdot c_z \cdot m \cdot \omega_0 \cdot M_2}{h_d \cdot D_{cp}^2 \cdot B \cdot \cos\alpha_w}, \text{ [m]}, \tag{7.2}$$

where k [1/Pa] is the coefficient of wear which depends of friction materials and of wear conditions; c_z is the number of teeth in coupling of gears expressed as a part of rigid gear teeth number; ω_0 is the relative size of flexible thin-wall barrel cylinder; M_2 [N·m] is the load expressed in the form of rotation moment; h_d is the relative depth of teeth coupling; D_{cp} [mm] is the mean diameter of gears coupling; B [mm] is the length of tooth; α_w [degree] is the angle of action (gearing).

With help of Equation (7.2) and with the help of wear parameters of rigid gear it is possible to calculate the longevity of harmonic gear rotary feedthrough. It must be taken into account that rigid gear is manufactured of steel AISI 420FSe (Russian analogue type 40X13) or of steel ASTM 1045 (Russian analogue "steel 45") which corresponds to components: $U_0 = 1.53 \cdot 10^{-10}$ m, $k = 2.26 \cdot 10^{-13}$ 1/Pa and that flexible gear is manufactured of steel AISI 321 H (Russian analogue type 12X18H10T) at components: $k = 7.9 \cdot 10^{-14}$ 1/Pa. In this case the longevity of harmonic gear rotary feedthrough may be calculated as

$$N = \frac{h}{U_0} = \frac{h \cdot h_d \cdot D_{cp}^2 \cdot B \cdot \cos\alpha_w}{16.3 \cdot k \cdot c_z \cdot m \cdot \omega_0 \cdot M_2}, \tag{7.3}$$

where h [mic] is the thickness of tooth coverage made of MoS_2.

The calculation which was done with the help of Equation (7.3) shows that in case of coverage thickness $h = 70$ mic the calculated longevity of rigid gear is $N = 1.63 \cdot 10^6$ revolutions and the calculated longevity of flexible gear is $N = 4.66 \cdot 10^6$ revolutions.

The longevity experimental control of 24 drives, which rigid teeth were done of steel AISI 420Fse (Russian analogue type 40X13) and were covered with cover thickness $h = 70$ mic and which flexible gear was done of steel AISI 321H (Russian analogue type 12X18H10T) shows good agreement of the experimental results with calculated results.

Table 7.2 shows the longevity coincidence of the drive designed by authors of this book (Variant III) with variants designed by other designers.

7.4 Outgassing Flow of Harmonic Rotary-Motion Feedthrough

The summarized outgassing flow (Q_Σ) which is evolved from harmonic rotary-motion feedthrough into vacuum chamber, may be expressed as

Table 7.2 Coincidence of designs of different drives longevity.

Variant	Reference	Longevity without load (revolutions)	Longevity with load (revolutions)
VII	[8]	–	5×10^5
VI	[7]	1.2×10^7	2×10^5
III	[3]	$> 10^7$	3.7×10^6

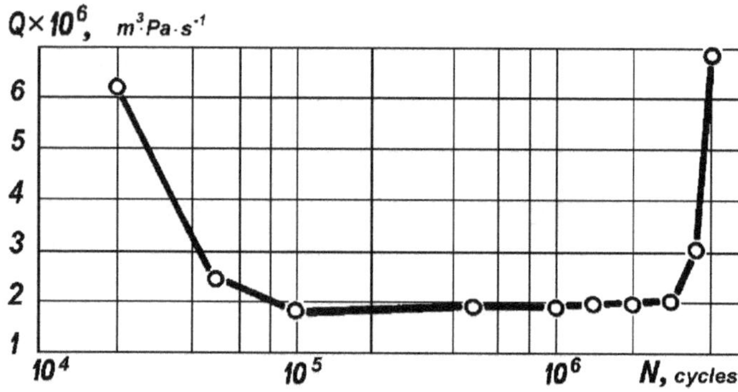

Fig. 7.3 Diagram of summarized outgassing flow variation during the life work period of harmonic rotary feedthrough (Variant III).

$$Q_\Sigma = Q_N + Q_D + Q_D + Q_F, \ [\mathrm{m}^3 \cdot \mathrm{Pa/s}], \tag{7.4}$$

where Q_N [m^3·Pa/s] is the gas flow, which penetrates through the cracks, welded seams and sealing elements; Q_D [m^3·Pa/s] is the flow of gases which desorbs from the feedthrough surfaces inverted into vacuum volume; Q_P [m^3·Pa/s] is the flow of gases which penetrate through the walls of the drive; Q_F [m^3·Pa/s] is the flow of gases which are generated in work process by friction couples of the drive (situated in vacuum).

When the drive stops at a temperature of up to 450°C, the outgassing flow consists mainly of gases which desorb from the surfaces of the drive (Q_D).

In case the drive is working, the outgassing flow increases because the component of gas flow appears from friction couples. This component appears mainly from coupled gears.

The gas flow mentioned may be many times (some orders) higher than gas flows released from other surfaces. Figure 7.3 shows the results of outgassing flow research from the gear couple of a rotary-motion feedthrough. The flexible gear of a couple was manufactured from steel AISI 321H (Russian analogue steel type 12X18H10T), the rigid gear of couple was manufactured from steel AISI 420 (Russian analogue steel type 40X13) and it was covered by a MoS_2 film with the help of the plasma method.

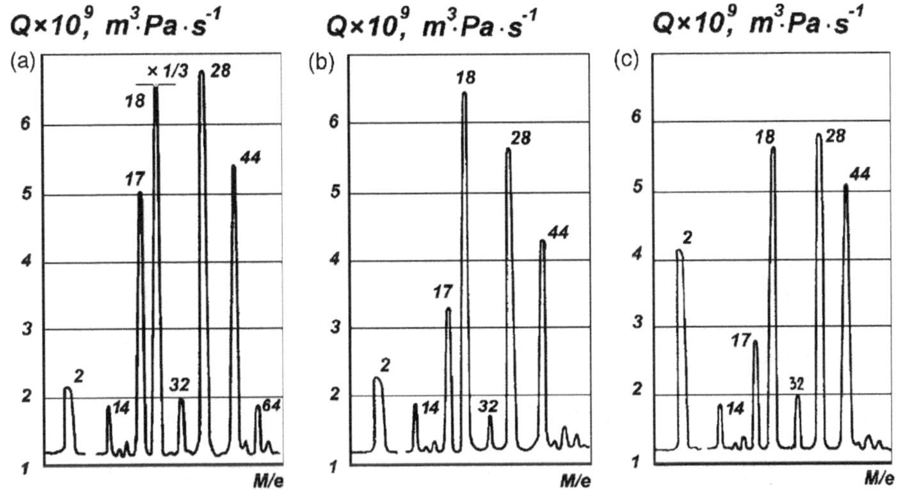

Fig. 7.4 Mass spectra of gases: (a) initial period at work, (b) stable work, (c) catastrophic wear.

The experimental conditions in the research were: the thickness of a rigid gear cover (film) $h = 70$ mic, diameter of flexible gear $d = 42$ mm, loading moment $M_2 = 5$ N·m, frequency of generator rotation $n_1 = 2500$ rev/min.

As we can see from the diagram presented, the friction process (in condition of coverage presence) may be divided into three periods which are different in character of outgassing flow variation and which are different in the value of the gas flow. In the initial period of work ($N = 0 \ldots 10^5$ cycles) the gas flow decreasing process (from $6.5 \cdot 10^{-6}$ to $1.8 \cdot 10^{-6}$ m^3·Pa/s) takes place. Then follows the longest period of stable work ($N = 1 \cdot 10^5 \ldots 3.7 \cdot 10^6$ cycles) and the outgassing flow is permanent and it is equal to $Q_F = (1.8 \ldots 2.1) \cdot 10^{-6}$ m^3·Pa/s in this period. Starting from $N = 3.7 \cdot 10^6$ cycles the outgassing flow begins to rise to a value of $Q_F = 4.6 \cdot 10^{-6}$ m^3·Pa/s.

Outgassing flow in the work process is characterized by the variation of the composition of gases, which are generated from friction couples. Figures 7.4a–b show the mass-spectra of gases which are generated in every above-mentioned period.

The signature of the breaking-in period ending (Figure 7.4a) is partial pressure of water H_2O (18^+) decreasing till the pressure which is close to nitrogen N^2 (28^+) pressure (Figure 7.4b). The signature of catastrophic wear of contacting zone period beginning (Figure 7.4c) is increasing of partial pressure of hydrogen H_2 (2^+) in result of adhesive wear beginning.

The flow of gases desorbed from contacting and near contacting surfaces depends on the temperature of these surfaces. The temperature of contacting surfaces depends on specific friction power which is generated on these surfaces, it depends also on contacting time, and depends also on conditions of heat conductivity. This specific friction power and as a result the outgassing flow of friction couples depends on contacting friction speed and depends on specific pressure. As a result of these conclusions the outgassing flow Q_F generated from contacting area of working couple of harmonic rotary feedthrough may be expressed by next equation:

$$Q_F = q \cdot F_{\text{cont}}\ [\text{m}^3 \cdot \text{Pa/s}], \tag{7.5}$$

where q is the outgassing flow expressed in reduced to contacting area unit form $[\text{m}^3 \cdot \text{Pa}/(\text{s} \cdot \text{mm}^2)]$; F_{cont} is the area of contacting surface $[\text{mm}^2]$.

The statistic transformation of high number of experiments help to create the expression for calculation the outgassing flow q which is coerced to contacting area unit in period of stabilized work of harmonic rotary motion feedthrough:

$$q = (0.023 \cdot p + 53.840 \cdot V_s + 1.070 \cdot p \cdot V_s \cdot 10^{-9},\ [\text{m}^3 \cdot \text{Pa}/(\text{s} \cdot \text{mm}^2)], \tag{7.6}$$

where P is the pressure in contacting points of teeth [MPa]; V_s is the sliding speed in teeth of gears couple [m/s].

Coefficient of variation V in case of statistic estimation of outgassing flow value may be calculated as follows:

$$V = \frac{\sigma(q)}{q_m} = 0.06, \tag{7.7}$$

where $\sigma(q)$ is the mean square deviation of measured flow value; q_m is the mean value of specific outgassing flow.

The analysis of well-known different types of harmonic dear rotary motion feedthrough showed that the designer's tendency to create a more manufacturable design led to negative results – the drives became heavier and their parameters became worse.

The optimal variant of the design was done because of the most rational form of flexible thin-wall flexible cylinder of barrel, also it was done because of special technological operations used in the process of its manufacturing.

Complex research of precision, longevity, outgassing flow, which were done for a large number of feedthroughs helped to create a reliability passport for the authors design of the drive and this design is of the best parameters in comparison with another known analogue designs.

Analysis of the outgassing flow from hermetic harmonic gear feedthrough shows that increased outgassing flow in the initial period of work is characterized by the next sources:

- outgassing flow from near contact surfaces;
- outgassing flow from oxidized films which arose in the period of work in atmosphere.

In the stabilized period of work the outgassing flow is characterized by the next sources:

- diffusion process from heated zones of feedthrough;
- gases desorption from friction zones of coverage.

After coverage wear (after $N = 3.7 \cdot 10^6$ cycles) the clean metal surfaces come into action and the process of adhesion wear begins.

In this period of work the outgassing flow increases because of intensive hydrogen detachment from the microparticles which are pulled out from the contacting surfaces and where hydrogen is dissolved in the process of adhesive wear. These microparticles contain increased quality of dissolved hydrogen. According to the theory [8] these microparticles are squeezed between surfaces in the zone of friction contact. These microparticles are loaded with highest contacting stresses, they are heated to the highest (in contacting zone) temperature and they detach the hydrogen dissoluted in metal most intensively. The physical explanation of this process is done in Chapter 3.

Thus the friction process in case of coverage usage, may be divided into three periods:

1. Friction couple burn-in process, which is accompanied by dissolved gases elimination, adsorbed gases elimination, oxidized films of contacting surfaces and oxidized films of near contacting zones elimination.
2. Stationary process of coverage wear which is accompanied with constant in time outgassing flow.
3. Adhesive wear process, which is accompanied by increased outgassing flow which consists mainly of hydrogen.

7.5 Calculation of Hermetic Harmonic Gear Feedthrough

Let us consider the example of calculation of ultrahigh vacuum harmonic gear rotary feedthrough.

Initial data:

1. torsion moment of output shaft (kg·sm), $M_t = 250$ kg·sm;
2. transmission ratio of harmonic rotary feedthrough, $u = 101$;
3. maximum manufacturability of design;
4. longevity (revolutions), $N > 10^5$ revolutions;
5. kinematic precision, $\Delta\varphi = 30$ angular sec. ($\pm 15''$).

We select Variant III of flexible gear, which ensures the optimal correlation of parameter values in accordance with Table 7.1.

7.5.1 Determination of the Number of Teeth

A two-disc wave generator is used in the drive. The generator creates 2 waves of flexible gear deformation and according to Equation (7.8) we receive the correlation between teeth and wave numbers:

$$z_C - z_F = U \cdot K_Z, \tag{7.8}$$

where z_C is the teeth number of rigid gear; z_F is the teeth number of flexible gear; U is the number of deformation waves of flexible gear; K_Z is the teeth number difference in respect to one wave of flexible gear.

In our example $K_Z = 1$; $U = 2$. In this case, Equation (7.9) helps to calculate the teeth number of rigid gear z_C:

$$z_C = U \cdot u = 2 \cdot 101 = 202. \tag{7.9}$$

In this case Equation (7.10) helps to calculate the teeth number of flexible gear:

$$z_F = z_C - u = 202 - 2 = 200. \tag{7.10}$$

With the help of Equation (7.11) we can calculate the diameter of flexible gear:

$$d_F = \sqrt[3]{\frac{10 \cdot M_t}{\psi_d [\sigma_C]}} = \sqrt[3]{\frac{10 \cdot 250}{(0.1 \ldots 0.2) \cdot 90}} \cong 6.0 \text{ cm}, \tag{7.11}$$

where $\psi_d = b/d = 0.1 \ldots 0.3$ is the coefficient of gear disk rim tread (b is the gear disk rim tread); d is the outer diameter of thin-wall cylinder jacket; σ_C is the contortion breaking point.

7.5.2 Calculation of Main Sizes of Flexible Gears

Module of gears in gearing of rotation feedthrough drive may be calculated with the help of Equation (7.12):

$$m = \frac{d_F}{z_F} = \frac{60}{200} = 0.3 \text{ mm}, \tag{7.12}$$

In compliance with standard (Russian Standard of USSR 9563-60) we choose module $m = 0.3$ mm. Then we specify dividing diameters of rigid (d_C) and of flexible (d_F) gears:

$$d_C = m \cdot z_C = 0.3 \cdot 202 = 60.6 \text{ mm},$$

$$d_F = m \cdot z_F = 0.3 \cdot 200 = 60 \text{ mm}.$$

Then we specify teeth length of gear disk rim tread (gear disk rim width) with the help of Equation (7.13):

$$b = \frac{10M_t}{[\sigma_C] \cdot d_F^2} = \frac{10 \cdot 250}{80 \cdot 6.0^2} \cong 0.9 \text{ cm}. \tag{7.13}$$

Gears of rotary-motion feedthrough manufacturing may be done with the help of traditional teeth cutting instrument which ensures evolute profile of being cut teeth. It is better to cut the flexible gear on arbor in non-deformed state.

The form of flexible gear deformation may be done with help of reference profile which corresponds to standard (Russian Standard 3058-54) which ensures ($\alpha_D = 20°$, $f = 1$, $C = 0.25$).

Then we specify dividing diameters of gears coupling also as we specify teeth width in dividing diameter as it is shown below.

Let us specify the gear teeth parameters:

- determination of coefficient of flexible gear displacement as a function of teeth number z_F:

$$\chi_F \cong 2.15 \cdot 0.009 z_F,$$

$$\chi_F \cong 2.15 \cdot 0.009 \cdot 200 = 3.95\ (\alpha_D = 20°);$$

- determination of coefficient of reference profile displacement of rigid gear:

$$\chi_C = \chi_F - 0.15,$$

$$\chi_C = 3.95 - 0.15 = 3.8.$$

Displacement of reference profile:
Flexible gear $\chi_F = 3.95 \cdot 0.3 = 1.185$ mm;
Rigid gear $\chi_C = 3.8 \cdot 0.3 = 1.14$ mm.

Diameter of dedendum circle:
Flexible gear $d_{fF} = 0.3(200 - 2.1 - 2 \cdot 0.25 + 2 \cdot 3.95) = 61.62$ mm;
Rigid gear $D_{TC} \geq 62.67 + 0.54 + 0.195 = 62.96$ mm.

Diameter of addendum circle:
Flexible gear $d_{aF} = 61.62 + 3.5 \cdot 0.3 = 62.67$ mm;
Rigid gear $d_{aC} = 61.62 + 0.54 \cdot 0.195 = 62.355$ mm.

Flexible gear teeth are cut with help of hob on the gear-hobbing machine and rigid gear teeth are cut on the gear-shaping machine.

7.5.3 Calculation of Control Rollers Size of Rigid Gear

Let us calculate the control size using Equation (7.14):

$$M_C = mz_C \frac{\cos \alpha_D}{\cos \alpha_R} - d_P, \text{ mm} \tag{7.14}$$

where α_D is the angle of thread of original profile (20°); α_R is the angle of pressure action in contacting point of roller (this angle may be found with the help of evolute function tables); d_P is the diameter of control rollers $d_P = 0.572$ mm, rollers diameter in case of module size, $m = 0.3$ mm In this case:

$$m = 0.3\ \text{mm}, \quad z_2 = 202, \quad \text{inv}\,20^\circ = 0.014904,$$

$$\text{inv}\,\alpha_R = \frac{\pi}{z_C} - \frac{d_P}{d_{OC}} + \text{inv}\,\alpha_D - \frac{S_W}{d_C}, \tag{7.15}$$

where S_W is the tooth root width of rigid gear, mm

$$\text{inv}\,\alpha = \frac{3.14116}{202} - \frac{0.572}{56.9452} + 0.014904 - \frac{0.35868}{60.6} = 0.026319.$$

Therefore $\alpha_R = 24^\circ$.

In this case

$$M_C = 0.3 \cdot 202 \frac{0.9397}{0.9135} - 0.572 = 61.765.$$

Comment 1. After toothing geometry calculation and after flexible gear strength analytical control we control gearing quality, and then we control value of gaps and absence of teeth interference.

7.5.4 Calculation of Flexible Gear Geometry, Calculation of Geometry Sizes which Ensure Hermetic Properties of Flexible Gear

In correspondence with described method in case of ultrahigh vacuum feedthrough design we choose stainless steel type AISI1321H (1X18H10T, Russian standard 5582-61) as a base material for jacket of thin-wall flexible gears manufacturing.

The length L of thin-wall flexible gear may be determined in correspondence with correlation (7.16):

$$L = (2 \div 2.5) d_F\ \text{mm},$$

$$L = 2.1 \cdot 60 = 126\ \text{mm}. \tag{7.16}$$

The thickness h of flexible gear disk rim may be determined with Equation (7.17):

$$h = \left(75 + \frac{z}{4}\right) d_F \cdot 10^{-4}\ \text{mm},$$

$$h = \left(75 + \frac{200}{4}\right) \cdot 60 \cdot 10^{-4} = 0.75\ \text{mm}. \tag{7.17}$$

At that time we have the next correlations (in correspondence with Equation (7.18)) of main sizes:

$$h < 0.018 d_F \text{ mm.}$$

$$h < 0.018 \cdot 60 = 1.08 \text{ mm.} \tag{7.18}$$

In our case we choose

$$h = 0.81 \text{ mm.}$$

The thickness h_0 of thin-wall flexible gear disk rim may be determined in correspondence with correlation (7.19):

$$h_0 = (0.003 \div 0.005) d_F \text{ mm,}$$

$$h_0 = (0.003 \div 0.005) \cdot 60 = 0.18 \div 0.30 \text{ mm.} \tag{7.19}$$

Then we choose the final diameter sizes of gear:

$$h_0 = 0.23 \text{ mm}, \quad d_1 = 84 \text{ mm}, \quad d_2 = 66 \text{ mm.}$$

The calculation results of gear radial transference ω may be done in correspondence with (7.20):

$$\omega = 0.9 \cdot m = 0.9 \cdot 0.3 = 0.27. \tag{7.20}$$

In this case teeth size coefficient $Y = 0.648$. The greatest stress may be determined with the help of expressions (7.21) in (7.22):

$$\sigma_y \cong \frac{K_H \cdot C_\sigma}{Y} \cdot \frac{\omega \cdot h_e \cdot E}{r^2}, \tag{7.21}$$

$$\tau_y \cong \frac{M_t}{K_t \cdot 2\pi \cdot r^2 \cdot h_W}, \tag{7.22}$$

where K_H is the distortion coefficient of elastically distorted waveform under load, $K_H = 1.1 \div 1.4$; C_σ is the coefficient of distorted waveform of flexible gear disk rim; Y is the spanning angle of generator by the flexible gear disk rim; E is the modulus of elasticity, kg/mm^2; r is the radius of middle surface of thin-wall jacket of thin-wall flexible gears, mm; K_t is the coefficient of irregularity of torsion couple transformation (approximately $K_t = 0.2 \div 0.4$); h_{KP} is the thickness of weak section ($h_t = h_0$ for thin-wall jacket, $h_W = h_e$), mm; h_e is the equivalent thickness of thin-wall jacket rim of flexible gear which is calculated taking into account the high of gear teeth (in our case we can take into account $h_e = h_W + 0.6m$), mm h_r is the thickness of rim (thin-wall jacket rim of flexible gear under hollow), $h_r = h$. Taking into account:

$$\gamma = 40; \quad C_\sigma = 1.4; \quad r = 30.4; \quad K_H = 1.2; \quad K_t = 0.2.$$

We come to these expressions:

$$\sigma_y = \frac{1.2 \cdot 1.4}{0.648} \cdot \frac{0.27 \cdot 0.81 \cdot 10^4}{30.4^2} = 11.7\ \text{kg/mm}^2,$$

$$\tau_y = \frac{250}{0.2 \cdot 2 \cdot 3.14 \cdot 3.04^2 \cdot 0.081} = 280\ \text{kg/cm}^2 = 2.8\ \text{kg/mm}^2.$$

Safety margin stabilization must be done in flexible gear.

$$n = \frac{n_\sigma \cdot n_\tau}{\sqrt{n_\sigma^2 + n_\tau^2}} \geq 1.4, \tag{7.23}$$

where

$$n_\sigma = \frac{\sigma_{-1}}{K_\sigma \cdot \sigma_a}, \tag{7.24}$$

$$n_\tau = \frac{\tau_{-1}}{K_\tau \cdot \tau_a + 0.1\tau_m}, \tag{7.25}$$

σ_a, τ_a are the amplitude bending and shearing stresses, kg/mm^2.

The approximate fatigue limits are:

$$\sigma_{-1} = (0.4 \div 0.5)\sigma,$$
$$\tau_{-1} = (0.2 \div 0.3)\sigma.$$

Coefficients of efforts:

$$K_\sigma = 1.7 \div 2.0,$$
$$K_\tau = 0.7 \div 0.8.$$

It may be taken into account that the extreme amplitude stresses which act in flexible gear are as follows:

$$\sigma_a = \sigma_y,$$
$$\tau_a = \tau_m = 0.5\tau_y.$$

According to Equations (7.24) and (7.25) and taking into account $\sigma_{-1} = 28.5$ kg/mm^2, $\tau_1 = 17.1$ kg/mm^2, $K_\tau = 1.5$, $K_\sigma = 1.7$:

$$n_\sigma = \frac{28.5}{1.7 \cdot 11.7} = 1.46,$$

$$n_\tau = \frac{17.1}{1.5 \cdot 0.5 \cdot 2.8 + 0.1 \cdot 0.5 \cdot 2.8} = 7.6.$$

Safety margin stabilization according to Equation (7.23):

$$n = \frac{1.46 \cdot 7.6}{\sqrt{1.46^2 + 7.6^2}} = 1.43.$$

The safety margin is not very large, but nevertheless it is enough and good for this type of steel, because it has a good impact elasticity.

7.5.5 Calculation of Assurance Factor of Flexible Gear Teeth

As long as the length of the teeth of flexible gear is determined on the basis of Equations (7.16) and (7.11), which used σ_{CM} (contortion breaking point), it leads to the decision that no extra strength of the teeth is needed in addition.

7.5.6 Calculation of Flexible Gear Wave Generator

In the process of drive schemes analysis the two roller wave-generators were selected. The generator consists of two rollers with diameter D, which are based on ball bearings which in turn are based on an eccentric shaft. The principle of using two ball bearings (one ball bearing for every roller) helps decreasing warp in the rollers.

The rollers eccentricity e we can detect with help of Equation (7.26):

$$e = 3.4 \cdot m = 3.4 \cdot 0.3 = 1.02 \text{ mm}. \tag{7.26}$$

With the help of Equation (7.27) we can detect roller diameter:

$$D_P = d + 2\omega - 2e + \Delta = 60 + 0.54 - 2.04 + 0.09 = 58.59 \text{ mm}, \tag{7.27}$$

where Δ is the elastic deformation of ball bearing plus partial compensation of internal diameter of flexible gear. The value of parameter Δ for supporting ball bearing (Russian type No. 304) is accepted:

$$\Delta \cong 0.09 \text{ mm}.$$

The radial loading force P_R may be determined with the help of Equation (7.28):

$$P_R = 0.35m \cdot K \cdot h_2 \left(\frac{d_F}{L}\right)\left(\frac{d}{D}\right) \cdot E, \tag{7.28}$$

where K is the coefficient which depends on angle of thread of cutting instrument ($\alpha_D = 20°$; $K = 1$; $\alpha_D = 30°$; $K = 0.89$); d is the internal diameter of ball bearing: D is the outer diameter of ball bearing:

$$P_R = 0.35 \cdot 0.3 \cdot 1 \cdot 0.81 \cdot \left(\frac{60}{126}\right) \cdot \left(\frac{20}{52}\right) \cdot 10^4 = 155 \text{ kg}.$$

The radial loading force P_R helps to consider that complete loading force acts on one of the ball bearings of the roller. Equation (7.29) helps to calculate and to determine the permissible radial load of selected ball bearing (Russian type No. 304).

$$[R] = [\sigma^3]\frac{z \cdot d_b}{8720} = 280^3\frac{7 \cdot 9.53}{8720} = 170000\,\text{kg}, \tag{7.29}$$

where $[\sigma]$ is the permissible equivalent contacting stress of ball bearing $[\sigma] = 28000\,\text{kg/sm}^2$ which corresponds to longevity 10^9 revolutions in case of ball bearing work in the atmosphere; z is the balls number; d is the ball diameter, mm. As we can see the ball bearing corresponds to a condition of work which is: $P_R < R$.

References

1. *Harmonic Drive. Mechanical Power Transmission Systems*, Internal. Div. United Shoe Machinery Corp. Boston, MA, 1966.
2. *Perspectives of New Vacuum Equipment Manufacturing*, Book of Abstracts, All Russian Symposium, Moscow, 1980, issue 2, p. 128 [in Russian].
3. Deulin E.A., Methods of selection of reliable vacuum mechanisms structure, in *Book of Abstracts, IVC-12/ICSS-8*, The Hague, the Netherlands, 12–16 October, 1992, p. 430.
4. Mednikov, M.I., *Vacuum Motion Feedthrough*, Mashinebuilding, 1974, 182 pp. [in Russian].
5. Moori, A., *Moon Vehicle of Space Complex Apollo, Aspects of Space Technology*, Vol. 3, 1972, pp. 3–18 [in Russian].
6. De Haas E., The harmonic drive as UHV rotary motion feedthrough, *Nuclear Instruments and Methods* **137**, 1976, 435–439.
7. Hayashi Y., Yamada T., Development of the harmonic drive rotary feedthrough, *Vacuum* **41**(7–9), 1990, 1948–1950.
8. Deulin E.A., Concept of exchange of gases under friction in vacuum, in *Abstracts Book of 10th ICTF/5th EVC*, Salamanca, 23–27 September 1996, Madrid, 1996, p. 105.

Chapter 8
Ultrahigh Vacuum Non-Coaxial Linear-Motion Feedthroughs

Analysis of linear-motion feedthrough schemes (see Figure 4.2, Chapter 4) shows that sliding speed (V parameter) in kinematic pairs of feedthrough can vary from the maximum value corresponding to the traditional nut-screw couple to zero. The sliding speed variation from maximum value to minimum value leads to a great variation in the design schemes of the feedthroughs. In general, it makes the design of feedthroughs more complex.

One of the ways to decrease sliding speed is the use of bellows sealing elements. These elements directly transmit linear motion from the atmosphere into the vacuum chamber. Let us consider the mechanisms of linear motion which correspond to the upper line of Figure 4.2. In the mechanisms of this group, there is no friction in vacuum ($V_c = 0$). The simplest design of the mechanism consists of one bar sealed with a bellows [1]. The maximum length of the travel depends on allowable strain (stress) in the bellows [1] which can be determined from the following equation:

$$\Delta L = \frac{[\sigma] \cdot n \cdot (1 - \mu)^2 \cdot R_B^2}{E \cdot h \cdot k_{2B}},$$

where $[\sigma]$ is the allowable strain in the bellows; n is the number of crimps; μ is the Poisson coefficient; R_B is the internal radius of the bellows; k_{2B} is the coefficient which depends on the goffer depth.

When the length of linear travel is longer than the allowable deformation of one bellow, the bellows can be connected in series. In this case the travel of the bar is equal to the sum of the linear deformations of all bellows.

One of the disadvantages of these mechanisms is the large overall dimensions in combination with small linear transference in vacuum. In [2] the design of the mechanism based on a successive joint of bellows is described. This mechanism ensures linear transference of the bar of 120 mm. However, the overall dimensions of this drive are 9 times larger than for the single-bellows mechanism.

E.A. Deulin et al., *Mechanics and Physics of Precise Vacuum Mechanisms*,
Fluid Mechanics and Its Applications 91, DOI 10.1007/978-90-481-2520-3_8,

The linear-motion feedthroughs based on a chain of bellows have influence on the value of the residual pressure and on the composition of the residual atmosphere in vacuum. This influence depends on the leakage through the walls of the bellows and on the outgassing rate from the surfaces of the bellows. This problem limits the use of such mechanisms.

There is another type of linear-motion feedthrough in which the linear transference is not limited by the maximum allowable deformation of the bellows. This type of linear-motion feedthrough includes the impulsive feedthrough [3]. The impulsive feedthroughs use complicated mechanism on the atmospheric side of the drive having a large number of kinematic elements. Without these mechanisms it is impossible to ensure smooth movement inside the vacuum element, because of the interrupting (impulsive) movement of the stock. Thus, all of the metal mechanisms have the same problem. These mechanisms are either limited in value of transference, or they do not correspond to the required smoothness or uniformity of transference.

Bellows-sealed mechanisms of the second group (see Figure 4.2, Chapter 4) have small sliding speed. These mechanisms use rotating friction; then angular deformation of the bellow is transformed into linear motion of the bar. In this case the resulting transference is limited by the angular deformation of the bellows. Also, the resulting transference is not smooth.

Hermetic linear-motion harmonic drives shown in the left part of the second and third lines of Figure 4.2) provide the movement translation through the flexible hermetic thin wall which can have the form of a tube or a belt. The main disadvantage of these mechanisms is limitation of kinematic precision and the transmitted load because of the limited contacting stresses and elastic properties of thin-wall material.

The best variant for smooth movement situates on the fifth line (see Figure 4.2). The harmonic drives of this type are based on the combination of sliding and spinning friction. The disadvantage of harmonic nut-screw mechanisms is determined by the alternatingly varied deformation of the flexible nut. This is their main disadvantage. Another problem of the harmonic nut-screw mechanisms [2, 3] is the relatively low precision. This is because of large elastic deformations and large forces required to achieve these deformations. In addition, the intensive wear of nut and screw threads in vacuum [4, 5] limits the use of these mechanisms in ultrahigh vacuum equipment.

The mechanisms of linear movement of the sixth group (Figure 4.2, lower line) can be considered as a combination of a traditional nut-screw couple with traditional rotary-motion mechanisms. These combined mechanisms cannot be used in vacuum, because of too high sliding speeds. Traditional nut-screw mechanisms are very simple but their effectiveness is very low because of low output, intensive wear and high sliding speed in the thread. These factors lead to a high outgassing rate which decreases the efficiency of these mechanisms.

The higher efficiency can be obtained using rotating nut-screw couple. Using of the ball-screw helps to reduce friction forces, sliding speed, and to increase precision and smoothness of the transference. Nevertheless, these mechanisms have several disadvantages:

- High technological complexity and sensitivity to conditions of exploitation, including: using this mechanism in vacuum and outgassing heating.
- Inaccuracy in the manufacturing process of drive elements. It leads to decreasing the drive quality and functionality.
- The parameters of this type of drive depend on the value of preloading which is a function of load. Preloading leads to higher contacting stresses which is a reason of a higher wear in vacuum. The difficulty to remove the wear debries from contacting zone makes it impossible to use balls without liquid lubricant in the drive. However, using at liquid lubricants is not possible in many ultrahigh vacuum applications.

The group five of the table shown in Figure 4.2 offers the new possibilities for design of non-coaxial nut-screw drives and new types of these drives hermetization.

The principle of these mechanisms is based on transformation of rotary-motion into linear motion [6]. As shown in Figure 8.1 the nut 5 has a thread diameter larger than the screw 7. The nut is fixed to a ball bearing 4. Rotation of the eccentric ring 3 having an eccentric hole leads to nut thread running-in in coupling contact with screw thread. The eccentric ring 3 is based on rotating bearings 2 inside a body 1 which is hermetically joined to the bellows 6. The drive hermetization is realized using a chain bellows between the nut and the body. In this drive because of the planetary motion of the nut, the sliding speed and forces on the contacting threads are smaller than in traditional nut and screw under mutual rotation.

This leads to decreasing of friction force in comparison with traditional nut-screw couple decrease of the friction torque helps to increase the drive sensitivity and to increase smoothness of its motion under very low sliding speeds [7].

The known designs of hermetic drives based on non-coaxial nut-screw couples [8] show that non-coaxial mechanisms are very promising for usage in ultrahigh vacuum equipment and in aggressive environment.

Taking into account multi-variant possibility of non-coaxial drives design [9] and multi-variant possibilities of the drive composition, it is necessary to consider the principles of their new designs.

8.1 The Hermetic Drive Designs Principles Based on Non-Coaxial Nut-Screw Couples

The basis of hermetic non-coaxial feedthrough drives is the transformation of circular motion of one element (screw or nut) of planetary nut-screw drive linear motion of the output element (Figure 4.6). The principle of operation of this mechanism is shown in Figure 8.1. The difference in threads steps must be taken into account. When the threads steps are the same, the nut planetary motion does not lead to coupled linear motion of the screw. The simplest variant of such kind of scheme is the

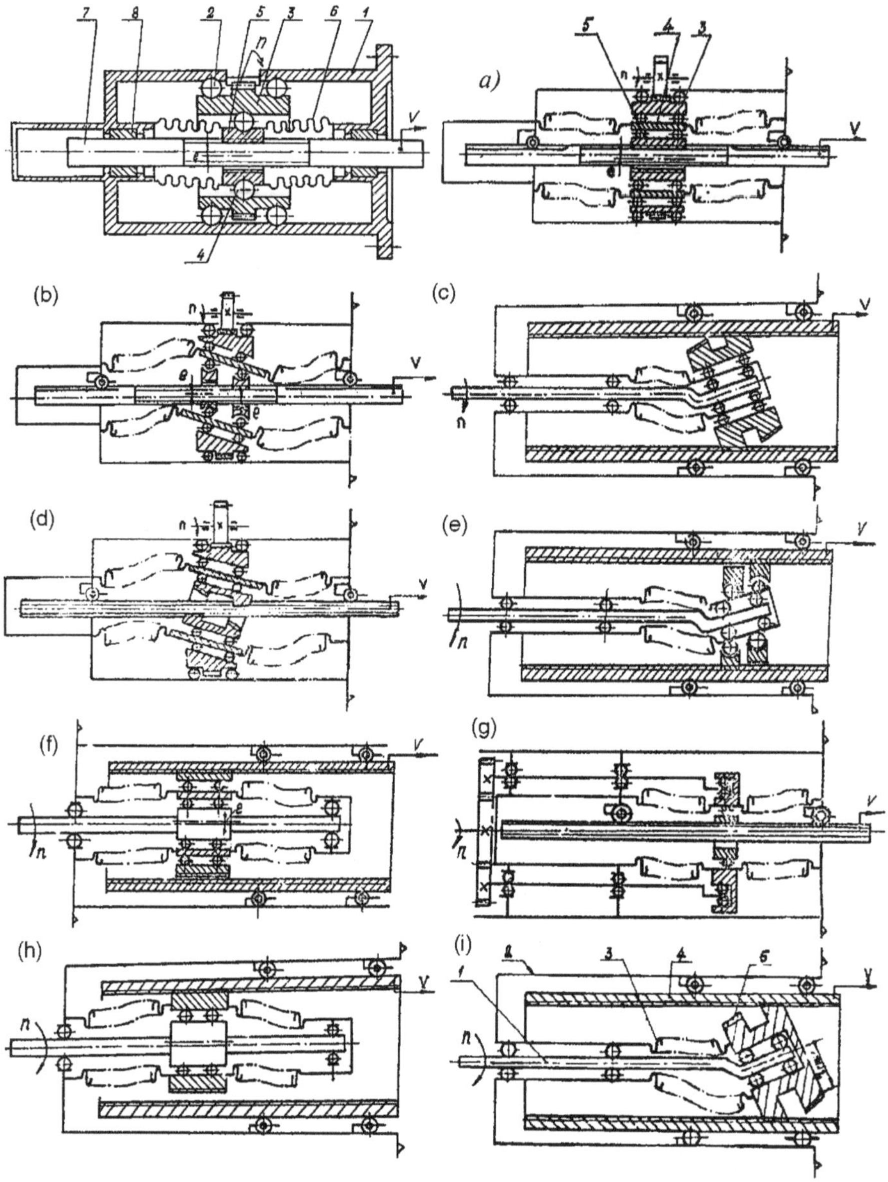

Fig. 8.1 Schemes of hermetic feedthroughs with different sliding speed in nut-screw coupling.

drive based on nut-screw coupling when the nut thread step is zero and the screw thread step is a unity.

The planetary nut-screw feedthroughs contains many friction elements situated in vacuum volume. These contacting elements are: ball bearings, thread couple,

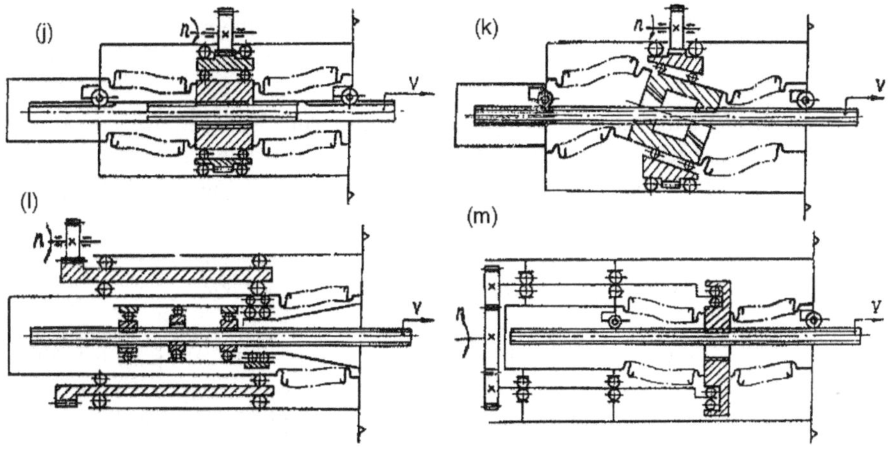

Fig. 8.1 Continued.

guides of threaded rod. The basic design element which determines the kinematic transformation of nut rotation into linear transference of the threaded rod is the thread couple. The workability of a planetary nut-screw feedthrough depends mainly on friction coefficient between the screw and the nut.

The design of hermetic nut-screw feedthroughs can be done on the basis of four basic schemes shown in Figures 8.1a, b. Drives designed according to these figures can have different sliding speed of coupling threads. The schemes shown in Figures 8.1b, d–g are the variations of above-mentioned schemes. In Figure 8.1 there are no schemes with fixed nut. This is because of increased friction and limited loading ability of this design in comparison with the drive with mothing nut.

The rolling nut-screw motion in the drive can be obtained by introducing additional intermediate unit (see pos. 4, Figure 8.1a). This unit is connected with the nut through the bearing 5. This unit is fixed in angular position using bellows welded to the unit. This unit moves in circular planar-parallel movement. It can be seen that the sealing bellows are welded only to intermediate unit but the bellows are not welded to the nut.

Let us consider the schemes of the drives and try to find the hermetically sealed mechanisms where the sliding speed is minimal. A low value of the sliding speed increases the longevity and reliability of the mechanisms. The value of the sliding speed in the thread couple can be determined according to the calculation method described in [9]. For single thread of the screw and the nut (see Figure 8.1) the nut can be considered as a tube with circular graves forming steps. The distance between graves is equal to the steps of the screw thread. In this case the sliding speed can be calculated as follows:

$$V_{CK} = \frac{\pi \cdot n}{60}\sqrt{(D_\Gamma - d_B)^2 + d_B^2 \cdot \tan^2 \alpha}, \qquad (8.1)$$

where n is the number of rotations of eccentric ring; D_Γ is the middle diameter of the nut thread; d_B is the middle diameter of the screw thread; α_B is the lead angle of screw thread.

If a pod bush 4 is introduced between the eccentric ring 3 and the nut 5 as shown in Figure 8.1 (see Figure 8.1a), the behaviour of friction in the thread is different: the rolling friction dominates and the sliding speed decreases four times as follows from the equation:

$$V_{CK} = \frac{\pi \cdot n}{30} \cdot r_B \sqrt{(1/\cos\alpha_B - 1)^2 + \tan^2\alpha_B}. \tag{8.2}$$

The third basic scheme is shown in Figure 8.1i. This scheme can be by using the inclined screw (Figure 8.1u) which is formed by two cones having circular graves. The graves form the thread with zero lead angle. The steps of the cones graves are equal to the thread steps of coreless threaded rod. This mechanism consists of the crank 1, the body 2, the bellows 3, the coreless threaded rod 4, and the nut 5. The sliding speed in the thread can be calculated according to

$$V_{CK} = \frac{\pi \cdot n}{30} \sqrt{(r_B - r_{\Gamma\,\min})^2 + r_B^2 \tan^2\alpha_B}, \tag{8.3}$$

where $r_{\Gamma\,\min}$ is the minimum average radius of the nut thread; r_B is the average radius of the screw thread.

The fourth basic scheme of the drive (see Figure 8.1c) can be formed by introducing a pod bush 6 in the scheme of the drive shown in Figure 8.1i. Sliding speed in the thread of this drive is expressed as

$$V_{CK} = \frac{\pi \cdot n}{15} \sqrt{r_{\Gamma\,\min}^2 + r_B^2 \tan^2\alpha_B} \cdot \sin\frac{\alpha_B}{2}. \tag{8.4}$$

The variants of the basic schemes help us to develop new designs of the nut-screw linear motion feedthroughs which have different numbers of elements of low reliability. The schemes shown in Figures 8.1e and 8.1f can be considered as modified variants of the schemes in Figures 8.1 and 8.1a. In these schemes the rod has the form of a tube with threaded internal surface. The nut has thread on the outer surface. The sliding speed for the scheme in Figure 8.1h can be determined from Equation (8.1). In the case of a scheme corresponding to Figure 8.1e can be determined with the help of Equation (8.2). The scheme corresponding to Figure 8.1l ensures low sliding speed and unloading of the screw bearings due to using three eccentric nuts. The scheme has a large number (7 as minimum) of ball bearings on vacuum side.

The schemes shown in Figures 8.1d and 8.1k can be considered as modified variants of the schemes shown in Figures 8.1c and 8.1h but the positions of the nut and the screw is different. The disadvantage of these schemes is the large number of sealing bellows required for the swinging nut (see Figure 8.1j) hermetization and the bush (see Figure 8.1k) hermetization. The advantage of these schemes is the simplicity of the screw design. The modification of the schemes shown in Figures 8.1c and 8.1d with the aim to decrease the sliding speed in the thread are shown in Figures 8.1b and 8.1e. The sliding speed in these schemes is determined by Equation (8.2). These advantages are obtained by increasing the number of elements on vacuum side. It should be taken into account that the nut of this drive is divided into two parts. The schemes shown in Figures 8.1g and 8.1m are modifications of the schemes shown in Figures 8.1a and 8.1b towards increasing of the sealing bellows reliability. The bellows in these schemes is protected from torsion in the case of a sudden inclination of the ball bearing of oscillating nut.

The classification matrix of nut-screw feedthroughs based on two criteria: (1) sliding speed in threads and (2) design of leading element (see Figure 4.6) reflects the basic schemes of planetary nut-screw feedthroughs. Further future development of those schemes for increasing of the loading ability increasing can be done in several ways:

- Using in feedthroughs special nuts. The threads of these nuts should be coupled with the threads of the screw and the eccenticities of the nuts should be mutually turned on a previously determined angle.
- Using in the feedthroughs special cone-ring nut or roller nut which should be inclined to the axis of the screw. The nut due to this inclination has two contact zones of the threads coupling.
- Using screws and nuts consisting of several rollers and coupling with screw. The nut as in the previous case moves in cross-plane circle oscillation movement due to its inclination.

8.2 Geometry of Nut-Screw Coupling of Linear-Motion Hermetic Feedthrough

One of the main questions which appears in the process of hermetic feedthroughs design is the determination of geometry parameters of nut and screw surfaces, for example, determination of the parameters of the nut surface which is coupled with screw surface. Both the internal and external coupling types can be used. The internal type is like in the case ring-nut schemes, while the external type is like in the case of schemes with rollers and outer nuts. The contacting zone can be considered as linear.

In general, calculation of the coupling geometry must consider selected mutually bending of spiral surfaces, cross-sections, curvatives, etc. So, from the point of view of differential geometry this task can be considered as a coupling of two spiral surfaces which axis are arbitrary oriented in the space. There is a number of methods for solutions of such tasks. The most useful method according to our opinion is a very simple and obvious kinematic method. This method is widely used for design of rotary-motion drives with screws. This method can be used also for creation of non-coaxial nut-screw couples. As the researches of non-coaxial couple show that

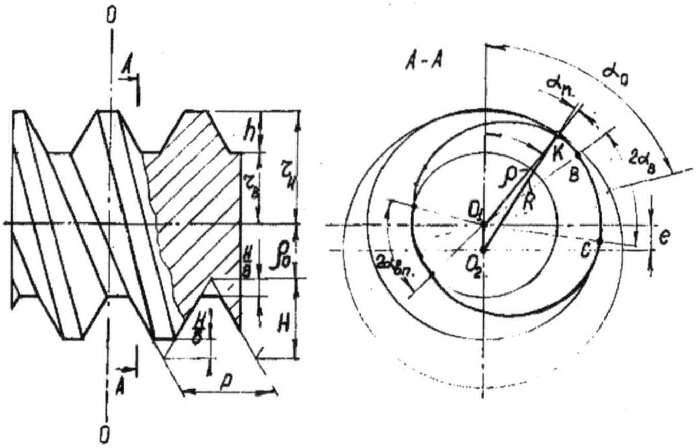

Fig. 8.2 Parameters of metrical screw.

the inclination of the screw profile for thread lead angle $\lambda \leq 5°$ is close to linear. This makes it possible to simplify the theoretic calculations as shown in [10, 11].

The parameters of the surface which must be coupled with another surface can be found using common methods which used in space or in plain toothing. At the same time, the process of hermetic non-coaxial nut-screw feedthroughs design can be improved using the following methods:

- It is necessary to choose a profile of designed spiral surface, which provides its linear contact with another screw surface.
- Inscribing the circle-screw profile into the profile obtained on the previous stage. The radius of the inscribed circle must to ensure good contacting rigidity and absence of contacting edge action in the process of the drive assembling. The described method is simple and ensures the high precision of calculations.
- In many cases, the task of determination of non-coaxial nut-screw feedthrough geometry consists only in determination of the nut thread profile which ensures contacting of this thread with the screw thread on the both sides of the thread profile.
- As an engineering method of circular nut-screw profile determination the following method can be recommended. The method of cross-sections can be used for circular nut-screw profile calculation. The metric screw-thread in front cross-section consists of two circle parts with radius r_H and r_B joined using Archimedean spiral parts as shown in Figure 8.2.
- The turn angle of symmetry axis of cross-section depends on the cross-section's position in respect to the screw axis and depends on the distance to the zero section according to

$$\alpha_0 = \frac{1}{P} 2 \cdot pi, \tag{8.5}$$

where P is the step of the screw.

The cross-section of the nut can be considered as a circle of radius R. Two cases of a contact between the screw and a circle nut can be considered: side type of contacting of outer edge of the screw thread with the nut (see contacting points B and C, Figure 8.2) and linear type of the screw and of the nut surfaces contacting. The last one determines the loading ability and kinematic precision of the most rigid drive.

Transformation of the turn angle α_0 of the screw axial cross-section corresponding to the side contacting type into surface contacting type between the screw and the nut can be expressed by the following equation:

$$\alpha_{01} = \arcsin\left[\frac{r_H}{l}\sin\alpha_{\Pi\,\min}\right] + \alpha_B + \alpha_{\Pi\,\min}, \tag{8.6}$$

where α_B is the angle of the lug circle (see Figure 8.2); $\alpha_{\Pi\,\min}$ is the minimum angle of Archimedean spiral angle of ascent (helix angle).

$$\tan\alpha_{\Pi\,\min} = \frac{H}{\pi \cdot r_H},$$

where H is the overall height of the thread profile (see Figure 8.2).

The radius of the nut determined by the side contact can be determined from:

$$R^2 = r_H^2 + e^2 + 2r_H \cdot l\cos(\alpha_0 - \alpha_B),$$

$$\alpha_B < \alpha_0 < \alpha_{01} < \pi - \alpha_B. \tag{8.7}$$

Equation (8.5) and Equation (8.7) allow us to determine the radius R of the nut. Side type of contacting in non-coaxial nut-screw gearing process can be avoided by increasing the nut radius as follows from Equation (8.7).

When the turn angle α_0 of the cross-section of the screw is larger than the angle α_{01} determined from Equation (8.6), there is a linear type of contact.

Formal use of the profiling methods for screw and cone surfaces profiling leads to transcendent equations. Profile calculation can be simplified using the following methods.

Let us consider the triangle O_1O_2K (Figure 8.2). Angle O_1O_2K has the apex in the point where the circle touches the Archimedean spiral. This angle is equal to the Archimedean spiral lead (helix) angle α_Π in the front cross-section of the screw. The radius R of the nut and its position on the axis 1 can be determined from ths triangle. This can be done as follows:

1. Assign the current radius of Archimedean spiral ρ beginning from the screw outer diameter r_H, then the apex angle α_Π can be determined according to

$$\tan\alpha_\Pi = \frac{H}{\pi \cdot \rho}. \tag{8.8}$$

2. The angle α of Archimedean spiral position (see Figure 8.2) can be determined from:

Table 8.1 Parameters of the nut and screw geometry.

Side contact on the screw lugs				Linear contacting						
6	6	6			6	5.5	5.2	5.05	5.08
			ρ	ρ	0.0802	0.0875	0.0925	0.09524	0.0947
				$\Rightarrow$					
				α_Π	0.58187	0.5888	0.59356	0.59615	0.5956
				R^*	6.8575	6.3559	6.05484	5.9042***	5.9344
0.359	0.5385	0.7181	α_0	α_0	0.97456	1.4292	2.64459	–	2.8954
0,1	0,15	0,2	$\Leftarrow$	ℓ	0.2714	0.398	0.7356	–	0.8064
7	6.991	6.955	R^{**}	R^{**}	6.8575	6.566	5.426	–	5.25
7	6.991	6.955	R	R	6.8575	6.3559	6.0548	–	5.9344

R^* – nut radius for linear contacting;
R^{**} – nut radius for side contacting when $e = 1$. The radius can be determined:

$$R^{**} = \sqrt{37 + 12 \cdot \cos\left(\alpha_0 - \frac{\pi}{8}\right)};$$

R – nut radius, $\max(R^*, R^{**})$.
The symbol $\Rightarrow$ marks the value used for calculation.
*** – nut radius smaller than $r_B + e = 5.9265$.

$$\sin(\alpha - \alpha_\Pi) \approx \frac{\rho}{\ell} \sin \alpha_\Pi. \tag{8.9}$$

3. The corresponding radius R of the nut can be calculated:

$$R = \ell \frac{\sin \alpha}{\sin \alpha_\Pi}. \tag{8.10}$$

4. Then the parameters α_0 and ℓ of radius R position are determined:

$$\alpha_0 = \pi - \alpha - \frac{(\rho - \rho_0)\pi}{H}, \quad \alpha_0 < \pi, \tag{8.11}$$

$$\ell = \alpha \frac{P}{2\pi}, \tag{8.12}$$

where ρ_0 is the minimum radius of Archimedean spiral for standard metric screw-thread:

$$\rho_0 = r_H - \frac{7}{8} H, \quad H = \frac{P}{2 \tan 30^\circ}. \tag{8.13}$$

For profile calculation it is necessary to assign the values of parameters and to determine values R and e. When ρ becomes close to r_B, the angle parameter becomes larger than π or the nut radius R is smaller than $r_B + e$.

With the iterative method of successive approximations it is possible to determine the minimum radius of the nut if the following limitations take place:

$$r_B + e < R, \quad \alpha_0 < \pi. \tag{8.14}$$

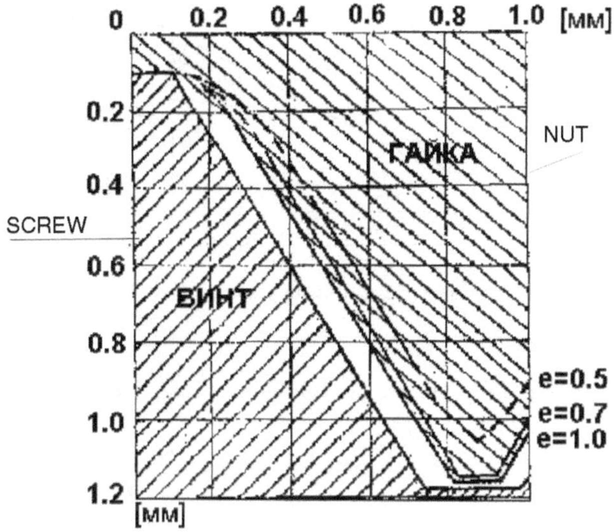

Fig. 8.3 Profile of circular nut for coupling with metrical screw M12x1.75.

In the case of small eccentricities, when $e < 0.5 \div 0.8h$ (h is the profile height, Figure 8.2) the process of nut thread profile fitting in the outer edge of the screw thread can be performed. The nut radius can be found from Equations (8.10) and Equation (8.7) and it corresponds to the maximum value of the radius. In our example, the profile of circle nut mated to the metric thread screw type M12X1,75 was determined for the cases of different eccentricities. The calculated profiles are shown in Figure 8.3.

The dashed line shows the parts of the nut determined by the side type of contacting, the continuous line shows the parts determined by the linear type of contacting. It can be seen from Figure 8.3 that decreasing of the eccentricity leads to decreasing of the length of linear contact. Table 8.1 shows the results of nut thread profile calculation for the eccentricity $e = 1$ mm.

8.3 Kinematic Calculation

In the kinematic calculations of nut-screw mechanisms the axial transference (or axial speed) of output element is determined as a function of axial transference (or axial speed) of the leading (of input) element. Taking into account that sliding speed in the nut-screw non-coaxial mechanisms is different from the sliding speed in traditional screw mechanisms, the kinematic calculation of the nut-screw non-coaxial mechanisms must be based on certain peculiarities.

In general, the axial transference (or axial speed) of the output element of the drive is a function of angular transference (or angular speed) in contacting points of

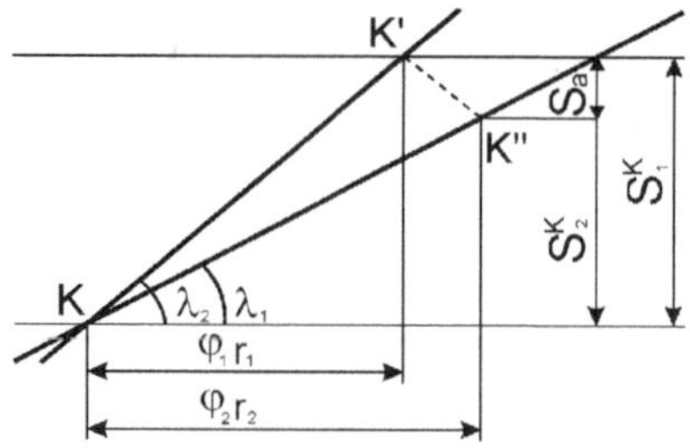

Fig. 8.4 The broaching draft (involute) of the screw and of the nut.

coupled screw surfaces. We can use the following equations:

$$S_a = S_1^K - S_2^K = P_1\varphi_1^K - P_2\varphi_2^K,$$
$$V_a = V_{a1}^K - V_{a2}^K = P_1\omega_1^K - P_2\omega_2^K, \qquad (8.15)$$

where φ_1^K, ω_1^K angular transference and angular sliding speed in direction of contacting screw surface of leading element of nut-screw couple; φ_2^K, ω_2^K is the angular transference and angular sliding speed in direction of contacting screw surface of driven element of nut-screw couple; P_1, P_2 is the screw parameters of coupled screw elements.

In traditional screw mechanisms which cannot be considered as non-coaxial mechanisms the angular transference and contacting sliding speeds along screw surfaces of coupled elements are equal but have different signs. It can be expressed in the following form:

$$\varphi_1^K = -\varphi_1, \quad \varphi_2^K = -\varphi_2,$$
$$\omega_1^K = -\omega_1, \quad \omega_2^K = -\omega_2.$$

Like traditional rotary-motion transmissions, the planetary non-coaxial nut-screw transmission can be reduced to the transmission with static axis. Taking into account the planetary type of the motion we obtain the following equations:

$$\omega_1^K = \omega_0 \quad \text{and} \quad \omega_2^K = \pm\omega_0 u_{21},$$
$$S_a = -P_1\varphi_0(1 \pm k \cdot u_{21}) \quad \text{and} \quad V_a = -P_1\omega_0(1 \pm k \cdot u_{21}), \qquad (8.16)$$

where φ_0, ω_0 are the angles of turn and angular speed of rotation of planet carrier; $k = P_2/P_1$ and $u_{21} = \omega_2/\omega_1$. The (+) sign corresponds to external coupling of the nut-screw contacting and the (–) sign corresponds to internal coupling. In engineering calculation, an approximate method of calculation can be used, which allows us to greatly simplify the calculations and to provide high precision.

The approximate approach of kinematic calculations is based on the assumption of the absence of slippage in the nut-screw coupling in the normal work. Let us consider the broaching draft of the screw and of the nut (see Figure 8.4).

As can be seen from the draft, the contacting point performs linear travel along the screw coil surface, while the leading element turns on angle φ_1. This linear way can be expressed as:

$$l^K = KK' = \varphi_1^K \frac{Z_{1cp}}{\cos\lambda_1}.$$

If there is no slip, linear travel can be expressed as:

$$KK'' = KK' = l^K.$$

The angular travel of the contacting point along the nut thread coil surface is not equal to the angular travel of the contacting point along the screw thread coil surface and it can be expressed as:

$$\varphi_2^K \neq \varphi_1^K.$$

Contacting point K passes in the axis direction the way along the screw surface h_1^K, and the way along the nut surface h_2^K.

As follows from the condition $KK' = KK''$ we can write the following equations:

$$h_1^K = \varphi_1^K \frac{r_{1cp}}{\cos\lambda_1}\sin\lambda_1, \quad h_2^K = \varphi_1^K \frac{r_{1cp}}{\cos\lambda_1}\sin\lambda_2.$$

In this case:

$$\begin{aligned} S_1 &= h_1^K - h_2^K = \varphi_1^K \frac{r_{1cp}}{\cos\lambda_1}\sin\lambda_1 - \varphi_1^K \frac{r_{1cp}}{\cos\lambda_1}\sin\lambda_2 \\ &= \varphi_1^K P_1 \left(1 - \frac{\sin\lambda_2}{\sin\lambda_1}\right). \end{aligned} \tag{8.17}$$

Taking into account that for the considered non-coaxial mechanisms $\varphi_1^K = -\varphi_1$ we can write:

$$S_a = \varphi_1 P_1 \left(1 - \frac{\sin\lambda_2}{\sin\lambda_1}\right).$$

Equation (8.17) is used only to the mechanisms which have no slippage. It means that it can be used for the mechanisms with only one degree of freedom.

The engineering calculation of axial transference and axial speed can be done using the following:

$$S_a = -P_1\varphi_1 \left(1 \pm \frac{\sin\lambda_2}{\sin\lambda_1}\right), \quad V_a = -P_1\omega_1 \left(1 \pm \frac{\sin\lambda_2}{\sin\lambda_1}\right). \tag{8.18}$$

When the parameter of circle nut of hermetic feedthrough $\lambda_2 = 0$, Equation (8.18) transforms into the form:

$$S_1 = P_1\varphi_1, \quad V_1 = P_1\omega_1.$$

8.4 Force Calculation of Hermetic Feedthroughs Based on Non-Coaxial Nut-Screw Mechanisms

The task of force calculation of non-coaxial nut-screw mechanisms consists in determination of contacting forces of coupled nut-screw mechanisms. Also it includes determination of torque, friction forces and efficiency of energy-conversion.

In contrast to traditional sliding nut-screw pairs, for the non-coaxial nut-screw pair at constant axial load the components of the load are not constant. These components depend on friction parameters in contacting points as well as in the bearings of the leaded element. It leads to variation of loads in bearing and other elements of mechanism loads. Also it can lead to variation of kinematic parameters of the mechanism. Thus, determination of loads between the elements of non-coaxial nut-screw transmission is one of the main tasks on the stage of the mechanism design. Real contacting spot of the nut-screw mechanism does not depend on theoretical contacting form: point or linear type of contacting. Usually the contact occurs in form of contacting spots. The shape and the sizes of the contacting spots depend on the following parameters: geometry of coupled screws, material properties, and applied load. Distribution of specific pressure in the zone of contacting area can be determined analytically. It depends on the coupling screws geometry parameters, manufacturing errors and mounting errors of the drive. The real contacting area is of a square shape. However, for simplification reason, in gear, worm, and other types of transmissions a point type of contacting is considered for the loads determination.

So, for the force calculation in non-coaxial nut-screw pair it can be assumed that the interaction force between a screw and a nut is concentrated in a point situated on the middle diameter of the screw. The friction force in a non-coaxial nut-screw pair consists of two components: sliding and rolling friction. In general, both of these components must be considered in the calculation. However, in some cases friction force component from rolling friction can be negligible. Direction of friction force vector F_f in the non-coaxial nut-screw mechanism depends not only on the leading element rotation direction but also on the variation of friction torque of output element. The vector of friction force varies direction in the region 2π (360°) in the plane normal to coupling surfaces. This variation is due to input and output threads directions rotation and resistance torque variation. Thus, the projection of the friction forces on coordinate axis can vary in value and sign.

As it can be seen from Figure 8.5, the friction force projected on the screw axis varies in the range:

$$-F_f < F_{fa} < F_f.$$

As a result, taking into account the friction force:

$$F_n^1 = F_a \frac{1}{\cos \alpha_\Pi \cdot \cos \lambda \pm f \cdot \sin \alpha_f},$$

where α_f is the angle between vector of friction force and the front plane. Then, the normal force F_n^1 can be considered as a variable even under constant axial load.

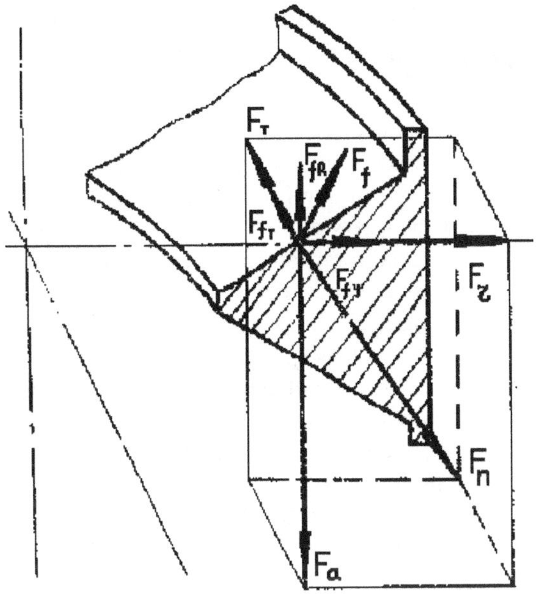

Fig. 8.5 Forces distribution at rotation in coupling zone of non-coaxial nut-screw pair.

The radial force which acts in coupling zone:

$$F_r^1 = F_n \cdot \sin\alpha_n \pm F_f \cdot \cos\alpha_f \cdot \sin\psi = F_a \left(\frac{\sin\alpha_n \pm f \cdot \cos\alpha_f \cdot \sin\psi}{\cos\alpha_n \cdot \cos\lambda \pm f \cdot \sin\alpha_f}\right)$$

$$= F_1 \left[\frac{\tan\alpha_n \pm \tan\rho' \cdot \cos\alpha + \sin\psi}{\cos\lambda \cdot \left(1 \pm \tan\rho' \cdot \frac{\sin\alpha_f}{\cos\lambda}\right)}\right],$$

where $\tan\rho' = f/\cos\alpha_n$; ψ is the angle between current radius-vector of contacting point and a perpendicular to the direction of friction force projection in the considered element of the mechanism.

The tangential force in coupling zone can be expressed by

$$F_t = F_n^f \cdot \cos\alpha_n \cdot \sin\lambda \pm F_f \cdot \cos\alpha_f \cdot \cos\psi$$

$$= F_a \left(\frac{\cos\alpha_n \cdot \sin\lambda \pm f \cdot \cos\alpha_f \cdot \cos\psi}{\cos\alpha_n \cdot \cos\lambda \pm f \cdot \sin\alpha_f}\right)$$

$$= F_a \left[\frac{\sin\lambda + \tan\rho' \cdot \cos\alpha_f \cos\psi}{\cos\lambda \cdot \left(1 \pm \tan\rho' \cdot \frac{\sin\alpha_f}{\cos\lambda}\right)}\right].$$

8.5 System Losses and Efficiency Factor of Hermetic Feedthroughs Based on Non-Coaxial Nut-Screw Mechanisms

Non-coaxial nut-screw mechanisms can be considered as a friction mechanisms which has several types of losses: contacting rolling friction in the thread coils, coils sliding friction, coils kinematic sliding friction, thread coils elastic sliding friction, friction in the bearings of the screw and of the nut. The friction losses in bearings are not specified for non-coaxial screw and nut mechanisms and these losses can be determined using well-known methods of calculation for specified conditions of the designed mechanisms loading. Calculation method of the friction loss in the threads contact of non-coaxial mechanisms is specific for this type of mechanism.

The main types of losses in the nut-screw non-coaxial couple is friction loss on the threads surfaces in mutual contacting. Three types of friction can be distinguished: pure kinematic slip, elastic sliding and geometrical sliding.

In the slip of contacting screw surfaces in non-coaxial mechanism a gap between contacting surfaces can appear. It leads to breach of mechanism kinematic parameters and failure of the mechanism.Thus, slip must be forbidden in the well-designed mechanism.

As it can be seen from the force analysis, the stress between the screw thread surfaces and nut thread surfaces is proportional to the axial load and inversely proportional to cosine of the product of the profile angle and thread inclination angle. So, non-coaxial nut-screw mechanism is characterized by autoregulation of the force between coupling surfaces. Therefore, this kind of friction loss is not taken into account.

In case of nut fixed against rotation, this sliding motion is not zero, so in each turn of the input eccentric, the nut slides in negative direction and its travel is equal to the nut and the screw coils difference. So, the value of kinematic friction loss can be expressed by:

$$N_{sl}^{kin} = F_n^f \cdot f_{fr}^{sl} \cdot V_{sl}^{kin}.$$

The elastic sliding friction resultin from the deformation in tangential direction as well as from rolling friction is very small because of high rigidity of coupling surfaces.

The most difficult part of the non-coaxial nut-screw drive calculation is the method of geometrical friction energy determination. In rotary-motion drives of friction type the contacting zone can be considered as a point or linear contact situated in the plane which coincides with the axis of the coupled gears. In the non-coaxial nut-screw drives the theoretical contacting zone can be of linear or of point form. From the manufacturing and from the assembling points of view, in order to avoid side contact point contacting is preferable. The danger of undesirable contacting stress increasing can be avoided by increasing the specific curvative radius of increasing the number of contacting points (in case of internal contact of the thread).

The value of geometrical sliding friction loss in the case of point contact can be determined according to:

$$N_F^{ck} = F_f \cdot V_{12} = F_n \cdot f \cdot V_{12},$$

where V_{12} is the geometrical sliding friction speed resulting from the contacting point inclination to the interaxis plane. In general, this speed is determined from the force conditions on friction contact [10]. In engineering calculation when the nut is not fixed against rotation, this speed can be determined according to:

$$V_{12} = \frac{V_a}{\sin \alpha_{oc}}.$$

When the nut is fixed against rotation using bellows welded to the nut, this speed V_{12} can be determined according to:

$$V_{12} = \frac{\omega_0 \cdot a_w}{\cos \lambda}.$$

The rolling loss of mutual contacting of screw surfaces are similar to the loss of elastic hysteresis in the rolling friction of gears couples and it can be determined by the following equation:

$$\begin{aligned} N^{ka} &= F_n \cdot k_p \cdot \omega_1 \cdot \sin \alpha_n \pm F_n \cdot k_p \cdot \omega_2 \cdot \sin \alpha_n \\ &= F_n \cdot k_p(\omega_1 + \omega_2) \cdot \sin \alpha_n \\ &= F_a \cdot k_p \frac{\omega \cdot \sin \alpha_n}{\cos \alpha_n \cdot \cos \lambda \pm f \cdot \sin \alpha_f}. \end{aligned}$$

In the case of hardened steels rolling friction coefficient is equal to 0.005. Thus, the total loss in non-coaxial nut-screw mechanism can be expressed as:

$$N_{pot} = N_f^{sl} + N^{ka} + \sum_{i=1}^{n} N_{op_i} + N_{sl}^{kin}.$$

The efficiency factor can be expressed as:

$$\eta = \frac{N_{Fa}}{N_{Fa} + N_{pot}} = \frac{1}{1 + \psi_{pot}},$$

where $\psi_{pot} = N_{pot}/N_{Fa}$ is the total loss coefficient:

$$\psi_{pot} = \psi_{sl}^{geom} + \psi_{ka} + \psi_{op} + \psi_{sl}^{kin}.$$

Of greatest significance is the loss of geometrical sliding of screw surfaces. For the mechanisms which have similar lead angles of threads of leading and output drive elements rolling thread loss can be close to the geometrical sliding loss.

For the efficiency factor determination it is necessary to calculate the value of the total friction loss coefficient in the mechanism. It should be taken into account that in the non-coaxial nut-screw mechanisms the total leading friction torque in the screw consisting of friction force of geometrical sliding, and rolling friction is opposite in sign and equal in value to all other kinds of leading element torques. So, this torque is a leading torque for the output element.

The total loss coefficient in non-coaxial nut-screw mechanism can be expressed as:

$$\psi_{pot} = \psi_{gs} + \psi_{ka} + \psi_{op1},$$

where ψ_{gs} and ψ_{ka} are the loss coefficients of friction, geometry sliding and rolling friction in nut-screw pair; ψ_{op1} is the friction loss coefficient in bearings of leading element.

Then the values of corresponding loss coefficients can be determined:

$$\begin{aligned}
\psi_{gs} &= \frac{N_{gs}}{N_{pc}} = \frac{F_f \cdot V_{12}}{F_a \cdot V_a} = \frac{F_a \cdot f \cdot V_a}{F_a \cdot V_a \cdot \sin\alpha_f(\cos\alpha_n \cos\lambda - f_a \cdot \sin\alpha_f)} \\
&= \frac{f_{pr}}{\sin\alpha_f \cdot \cos\lambda \left(1 - f_{pr}\frac{\sin\alpha_f}{\cos\lambda}\right)}; \\
\psi_{ka} &= \frac{N_{ka}}{N_{pc}} = \frac{F_a \cdot k_p \cdot \omega_{rel} \cdot \sin\alpha_n}{F_a \cdot V_a(\cos\alpha_n \cos\lambda - f \cdot \sin\alpha_f)} \\
&= k_p \frac{\omega_{rel} \cdot \sin\alpha_n}{V_a(\cos\lambda - f_{pr} \sin\alpha_f)}.
\end{aligned}$$

In the case when $P_1 = P_2 = P$,

$$\omega_{rel} = \frac{V_a}{P} = \frac{V_a}{r_{1cp} \cdot \tan\lambda}.$$

In this case:

$$\psi_{ka} = \frac{k_p}{r_{1cp}} \cdot \frac{1}{\left(1 - f_{pr}\frac{\sin\alpha_f}{\cos\lambda}\right)\cos\alpha_n \sin\lambda}; \quad \psi_{op1} = \frac{N_{op1}}{N_{pc}} = f_{op1}\frac{\omega_1 r_{op1}}{V_a}.$$

Then we can determine the efficiency factor:

$$r_{ox} = \frac{1}{1 + \frac{f_{pr}}{\sin\alpha_f \cdot \cos\lambda\left(1 - f_{pr}\frac{\sin\alpha_f}{\cos\lambda}\right)} + \frac{k_p \cdot \omega_{rel}}{V_a(\cos\lambda - f_{pr}\sin\alpha_f)} + \frac{f_{op1} \cdot \omega_1 r_{op1}}{V_a}}.$$

When the nut is fixed against the rotation it is necessary to supplement the efficiency factor with the kinematic sliding loss coefficient as it was shown above. The efficiency factor equation in this case is the following:

$$\psi_{kc} = \frac{N_{kp}}{N_{fa}}.$$

The efficiency factor in general case of the mechanism negative movement under load action may be written as

$$\eta_{ox} = \frac{N_{pn} - N_{pot}}{N_{pn}} = 1 - \psi_{pot}$$

$$= 1 - \left[\frac{f_{pr}}{\sin\alpha_f \cos\lambda \left(1 - f_{pr}\frac{\sin\alpha_f}{\cos\lambda}\right)} + \frac{k_p \cdot \omega_{rel}}{V_a(\cos\lambda - f_{pr}\sin\alpha_f)} + \frac{f_{op1} \cdot \omega_1 \cdot r_{op1}}{V_a} \right],$$

where N is the total supplied power.

8.6 Analysis of Loading Ability of Planetary Nut-Screw Feedthroughs

The loading ability of hermetic feedthrough is determined by the loading ability of non-coaxial nut-screw couple. The basis criterion of a non-coaxial couple workability is endurance of the contact screw work surfaces. Durability (strength) calculation of non-coaxial nut-screw couples can be done in general on the basis of contacting stress calculation in the same way as for friction rotation drives. The value of maximum contacting stress on the surface in case of linear contact of a nut and of a screw steel coils can be determined from the Hertz–Beliaev equation:

$$\sigma_N = 0.418\sqrt{\frac{q \cdot E_{pr}}{\rho_{pr}}},$$

where q is the normal specific load on a coil; E_{pr} is the effective elasticity module of nut-screw couple materials; ρ_{pr} is the reduced form of the curvature radius.

The normal specific load on a coil can be determined from

$$q = \frac{K_N \cdot F_H}{Z \cdot l_k} = \frac{K_N \cdot F_A}{Z \cdot l_k \cdot \cos\lambda \cdot \cos\alpha},$$

where Z is the number of nut coils; l_k is the length of contacting line in the coupling zone; K_N is the coefficient of the load (this coefficient takes into account the irregularity distribution of the load in the coils and distribution of the load along the contacting line). Load coefficient: $K_N = K_\beta \cdot K_Z$, where K_β is the coefficient of the irregularity distribution of the load along the contacting line; K_Z is the coefficient of the irregularity distribution of the load in the coils. In case of roller nut it is necessary to take into account the irregularity of the load distribution between the rolls. In this case the load coefficient can be determined as

$$K_N = K_\beta \cdot K_z \cdot K_m,$$

where K_m is the coefficient which takes into account the irregularity of the load distribution between the rolls.

Since the length of contacting line is small, the following equation can be used:

$$K_\beta = 1.$$

The results of theoretical and experimental research show that the load distribution between rolls and coils in the non-coaxial nut-screw couple depends on the design of the drive and also on the sizes of the coupling elements. In general, this distribution is similar to the load distribution in traditional nut-screw couples. So, the irregularity of the load distribution increases as a function of coils number in the coupling zone.

The precise specific value of the curvature radius of screw coil surface can be calculated using the quadratic form. However, this method is very complex for engineering calculations.

If the condition $\lambda \le 5°$ is satisfied, the curvatures of linear screw surfaces can be calculated in a first approximation replacing them by cone surfaces. These surfaces form the inclination angle equal to the profile of calculated surface angle. The calculation error in result of the above conditions does not exceed 5–10%.

In a first approach, the calculation model can be reduced to the internal coupling of two cylinders with their radiuses of curvature in a plane normal to the contacting line.

Hence, we have

$$\rho_1 = \frac{r_{1cp}}{\sin\alpha_1}; \qquad \rho_2 = \frac{r_{2cp}}{\sin\alpha_2},$$

and

$$\rho_{pr} = \frac{r_{1cp} \cdot r_{2cp}}{(r_{2cp} \pm r_{1cp})\sin\alpha}.$$

As for the calculation of specific curvatures, the calculation of the exact length of contacting line is not a simple task.

Since the contacting line position on the helical coil is close to the radial position, it can be assumed that the length of contacting line is equal to the width of the thread coil: $l_k = b$.

In this case:

$$\sigma_N = 0.418\sqrt{\frac{2 \cdot K_N \cdot F_a \cdot F_b \cdot E_2(r_{2cp} \pm r_{1cp})\tan\alpha}{Z \cdot b \cdot (E_1 + E_2) r_{1cp} \cdot r_{2cp} \cdot \cos\lambda}}.$$

The contacting stress is determined from

$$\sigma_N = 0.245 \cdot n_\sigma \sqrt{\frac{K_N \cdot F_{oc} \cdot E_{pr}^2}{Z \cdot \rho_{pr}^2 \cdot \cos\lambda \cdot \cos\alpha}}\ \text{kg/cm}^2.$$

The large axis b and the small axis a of contacting region are determined from

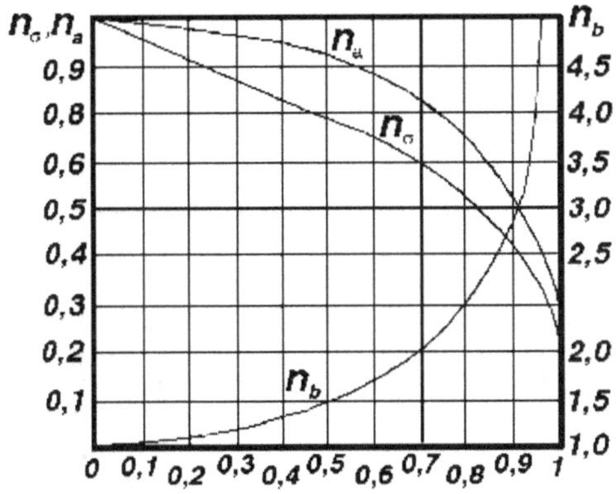

Fig. 8.6 Theoretical coefficients.

$$b = 2.8 \cdot n_b \sqrt{\frac{K_N \cdot F_{oc}}{E_{pr} \cdot Z \cdot \rho_{pr} \cdot \cos\lambda \cdot \cos\alpha}},$$

$$a = 2.8 \cdot n_a \sqrt{\frac{K_N \cdot F_{oc}}{E_{pr} \cdot Z \cdot \rho_{pr} \cdot \cos\lambda \cdot \cos\alpha}},$$

where n_σ, n_b, n_a are the theoretical coefficients which can be determined from the graphs (see Figure 8.6).

For determination of specific curvative radius it can be assumed that the thread of one coupling element of non-coaxial nut-screw is linear and the thread of another coupling element has small curvature. This is similar to the case of an axial section of the coil. Then,

$$\frac{1}{\rho_{pr}} = \left(\frac{\sin\alpha}{r_{1cp}} \pm \frac{\sin\alpha}{r_{2cp}} - \frac{1}{R}\right),$$

where R is the nut coil curvature radius in axial cross-section. The values of allowable contacting stress are shown in [9].

As a criterion of the hermetic feedthrough workability we can use the backlash of the screw coupling. This backlash appears as a result of wear of the nut and the screw surfaces.

Research of the backlash parameter of the screw coupling during wear of the nut-screw non-coaxial feedthrough shows that wear is uniform on the profile of the screw thread. Wear depends linearly on the life time as is shown in Figure 8.7.

The example of ultrahigh vacuum feedthrough based on non-coaxial nut-screw planetary drive was designed and successfully manufactured according to the described method and is shown in Figure 2.5b (see Chapter 2).

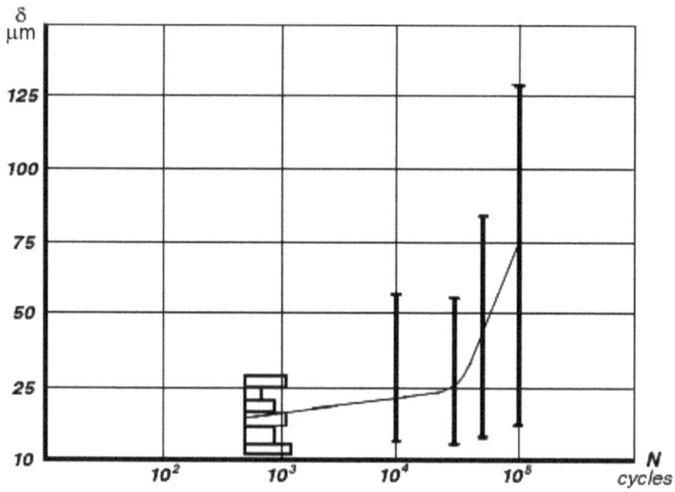

Fig. 8.7 Planetary nut-screw feedthrough wear process as a function of work (cycles number).

The main parameters of this drive are described below:

1. mounting flange: CF63;
2. length of the travel (feedthrough types: PNSE-350, 500, 750), mm: 350, 500, 750;
3. error of positioning, mm: 0.05;
4. maximum axial load, N: 300;
5. longevity, double cycles: 2×10^5;
6. non-coaxial nut-screw couple ensures small friction travel, small wear, high output efficiency;
7. unlimited length of the travel;
8. workability in any mouting position;
9. all metal-welded design, can be heated up to 450°C.

References

1. Khrunichev Y.A., Deulin E.A., Amosova E.P., *Design and Calculation of Vacuum Rotary-Motion Feedthroughs*, BMSTU, Moscow, 1977, 56 pp. [in Russian].
2. Mednikov M.I., *Vacuum Motion Feedthroughs*, Machinostroenie, Moscow, 1974, 184 pp. [in Russian].
3. Alexandrova A.T., *New Methods of Motion Generation and Transformation in Vacuum*, High School, Moscow, 1979, 69 pp. [in Russian].
4. Novoselov B.V., Bushenin D.V., *Mechanical Transmission of Servo-Drive Design* (Methodic Recommendations), Vladimir, 1980, 174 pp. [in Russian].
5. Pinegin S.V., *Contact Strength and Rolling Resistance*, Machinostroenie, Moscow, 1969, 243 pp. [in Russian].

6. Bushenin D.V., Method of rotation movement transformation into linear movement, Soviet Patent No. 271980, 1970 [in Russian].
7. Bushenin D.V., *Screw Mechanisms*, Vladimir, 1980, 72 pp. [in Russian].
8. Demande de Brevet d'Invention No. 7510525. Perfectionnements aux dispositifs de commande, notament de commande etanche d'un mouvement de translation par un mouvement de rotation, Roges Louis Elysel Germain. Date de depot 28 mars 1975.
9. Bushenin D.V., *Non-Coaxial Nut-Screw Drives* (Method of Design and Usage), Vladimir, 1974, 104 pp. [in Russian].
10. Marusov V.A., Mechanisms for linear travel automation in vacuum equipement, Automation of linear travel problems in electronics, in *BMSTY Proceedings*, Moscow, 1980, Vol. 334, pp. 119–129 [in Russian].
11. Bushenin D.V., E.A. Deulin E.A., et al., *Screw Mechanisms Design* (Methodic Recommendation), Vladimir, 1979, 128 pp. [in Russian].

Chapter 9
Vacuum Frictionless Mechanisms Based on the Principle of Controlled Elastic Deformation

The idea of these mechanisms was born between 1970–1980 in the former USSR and it was realized by Professor A.T. Alexandrova. The group of Professor Alexandrova designed and manufactured different vacuum manipulators, vacuum drives, vacuum rotary-motion and linear-motion feedthroughs, gates and valves based on the principle of controlled elastic deformation. This work was pioneer in the field of vacuum technique and technology.

In many cases the applicability of traditional gears, ball bearings and friction couples is limited in vacuum because of gas tribodesorption, wear of contacting surfaces, limited service life of friction couple (see Chapters 3 and 7). The solution of these problems can be found by controlling friction, wear and gas tribodesorption vacuum mechanisms as well as by use of frictionless mechanisms based on the principle of controlled elastic deformation.

Not all vacuum mechanisms could be realized using controlled elastic deformation of vacuum mechanisms. Nevertheless, in many cases, for example for fast-acting vacuum manipulators, transporting systems, precise positioning systems, bodiless vacuum gates and valves, and other small mechanisms, the use of controlled elastic deformation of vacuum mechanisms is very promising because of compactness, low weight and absence of interfacial friction.

The elemental basis of vacuum mechanisms without friction couples consists of various hermetic driving tube elements of different configurations (as shown in Figure 9.1). These elements can be deformed by supplying compressed gas or liquid inside the tube elements under controlled pressure.

The radius of curvature of the center axis of hermetic driving elements of different configurations can be varied according to different principles of variation. In case of non-finite number of elements (Figure 9.1a–g) it essentially influences the trajectory of free end of the driving element. The cross-section geometric shape of hermetic driving elements can be realized in the following variants shown in Figure 9.1j: elliptic, rhombic, concavo-concave and others. The geometric shape of the

E.A. Deulin et al., *Mechanics and Physics of Precise Vacuum Mechanisms*,
Fluid Mechanics and Its Applications 91, DOI 10.1007/978-90-481-2520-3_9,

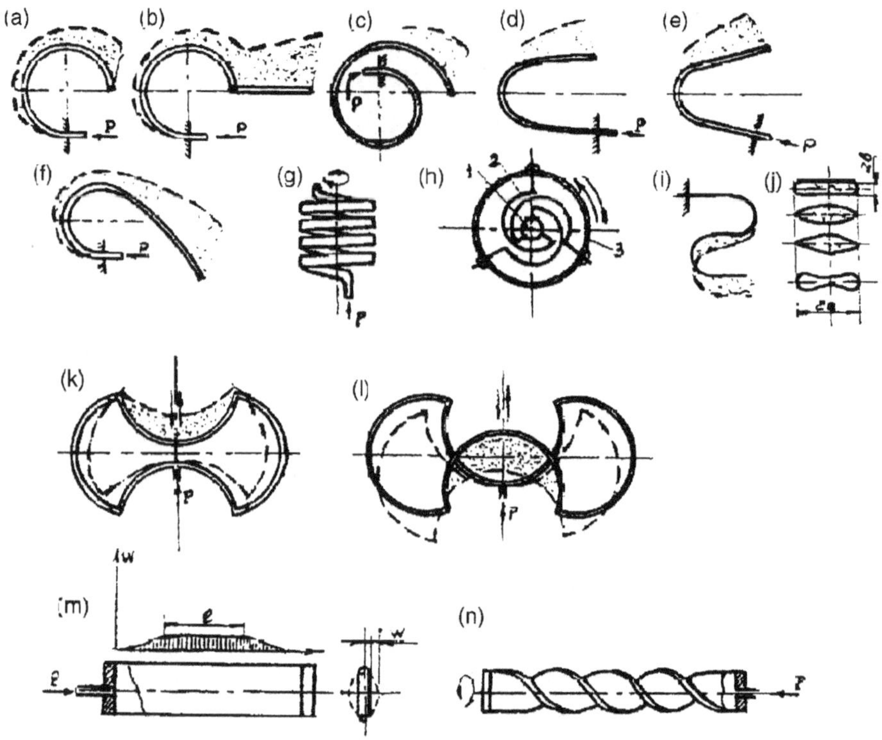

Fig. 9.1 Variants of the drives schemes based on controlled elastic deformation principle.

cross-section of hermetic driving element determines the following parameters of drives: sensitivity, maximum displacement and loading capability of the drive.

The simplest and the most technological types of design are the drives with constant radius of curvature (Figures 9.1a–b). The displacement range of these elements can be multiplied by their combination with layers.

Special drives with shape of Archimedes's spiral Figure 9.1c, parabola Figure 9.1d, sinusoid Figure 9.1e and cycloid Figure 9.1f are designed to realize displacement on specific trajectories.

Multi-turn can realize circular displacement for various angles which depend on the total number of coils. Circular displacement can be realized also using three free-ended drives (position 2 in Figure 9.1h). These elements are connected by one side to the supply line of gas or liquid by the collector 1, while another side is connected to the outer ring 3. The advantage of this scheme is the possibility of every drive 2 to ensure isolated radial output. The drives of closed contour (Figures 9.1k–l) are designed for linear displacement. The drives formed with linear axis are designed for precise linear displacement (see Figure 9.1m) and for precise angular (circular) displacements (see Figure 9.1n).

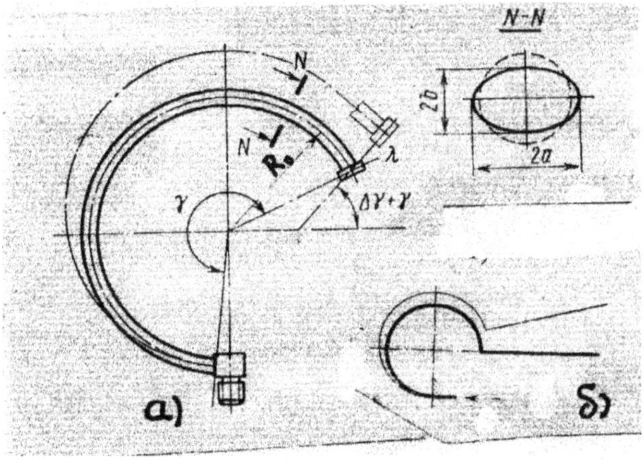

Fig. 9.2 Calculation scheme of the drive based on controlled elastic deformation principle which corresponds to open-ended contour.

New manipulators and mechanisms with a large number of degrees of freedom which can realize complex trajectories can be developed by combination of some of these driving elements of controlled elastic deformation either between them or with metal bellows.

Geometrical parameters of elastically deformed drives of free-ended type contour are shown in Figures 9.2a–b. The transference λ of the free end of this drive and developed force of this drive are determined using an energy method (Roots Method). In result of a second redevelopment of the Roots Method the following equations were obtained for simple engineering calculation:

$$\lambda = \frac{PR_0^3(D_6 + D_7\nu)}{Eah(D_3 + D_4\nu + D\nu^2)}\sqrt{(\gamma - \sin\gamma)^2 + (1 - \cos\gamma)^2}, \tag{9.1}$$

$$Q_\tau = 4Pa^2\frac{D_6 + D_7\nu}{D_0 + D_1\nu + D_2\nu^2}\frac{2(\gamma - \sin\gamma)}{3\gamma - 4\sin\gamma + 1/2\sin 2\gamma}, \tag{9.2}$$

$$Q_{r\tau} = 4Pa^2\frac{D_6 + D_7\nu}{D_0 + D_1\nu + D_2\nu^2}\frac{4(1 - \cos\gamma)}{2\gamma - \sin 2\gamma}, \tag{9.3}$$

where E is the modulus of elasticity of the drive material; Q_τ and Q_r are the tangential and radial tractive forces, respectively; R_0 is the initial radius of curvature of the central axis of the driving; h is the wall thickness of the element; a, b are the large and small axes of cross-section; $D_0 \ldots D_7$ are the coefficients which depend on the shape of normal cross-section and ratio of axes (see Table 9.1). The ratio of the large to small axes $a/b = k$; ν is the non-dimensional parameter which can be calculated from

$$\nu = \frac{R_0^2h^2}{12(1 - \mu^2)a^4}, \tag{9.4}$$

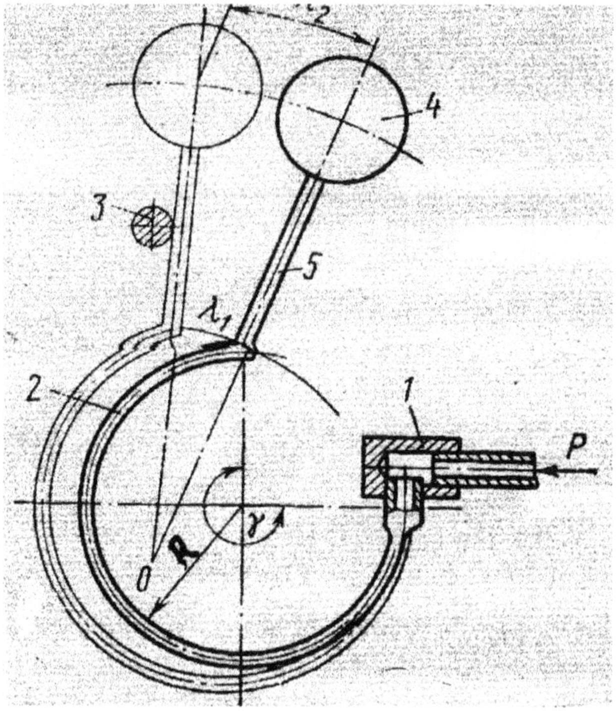

Fig. 9.3 Scheme of screening device.

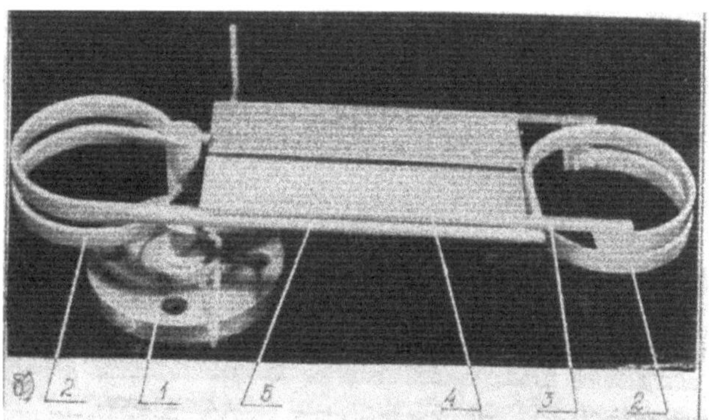

Fig. 9.4 Scheme of two-folding screening device.

where μ is the Poisson ratio.

One of the simplest devices based on controlled elastic deformation is the screening shield. Figure 9.3 shows the scheme of a screening shield which consists of the connecting pipe 1, the drive 2, the support 3 which limits the position of the screen 4, and the lever 5. Figure 9.4 shows a bivalve (butterfly-type) screening shield. The

Table 9.1 Table of coefficients.

View of cross-section	Ratio of large to small axes $a/b = k$	Values of coefficient for $h/b < 0.3$							
		D_0	D_1	D_2	D_3	D_4	D_5	D_6	D_7
Oval plane	2	34.5	33122	595992	0.7	1400	133010	9	9100
	3	78	65350	1295773	0.8	1503	133676	18	14256
	4	89	73692	1562383	0.6	1053	92410	17	12620
	5	89	73780	1643135	0.4	718	62904	14	10266
	6	85	71528	1654052	0.3	504	44302	12	8340
Rhombic $r = b/4$	1	91	75747	1854078	0.1	240	88603	16	11871
	2	71	62061	1613217	0.055	118	43063	9.5	7623
	3	62	55595	1495687	0.03	71	25540	7.0	5571
Rhombic $r = b/5$	1	86	72721	1816548	0.08	195	83185	14.5	11138
	2	67	59548	1576173	0.04	95	40173	9	7152
	3	58	53466	1461614	0.02	57	23777	6	5240

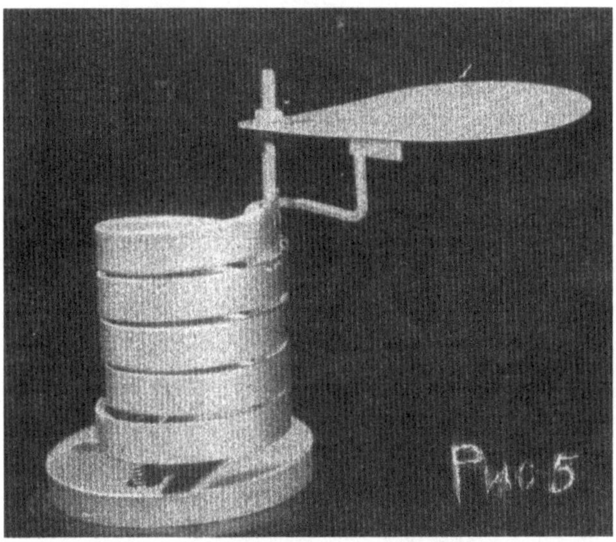

Fig. 9.5 Screening device with multi coil drive.

lever 5 moves parallel to the screen 4. The second pair of drives can be attached to the device in order to perform displacement in opposite directions (Figure 9.4). When the angular transference should be increased, a multi-turn drive can be used (Figure 9.5).

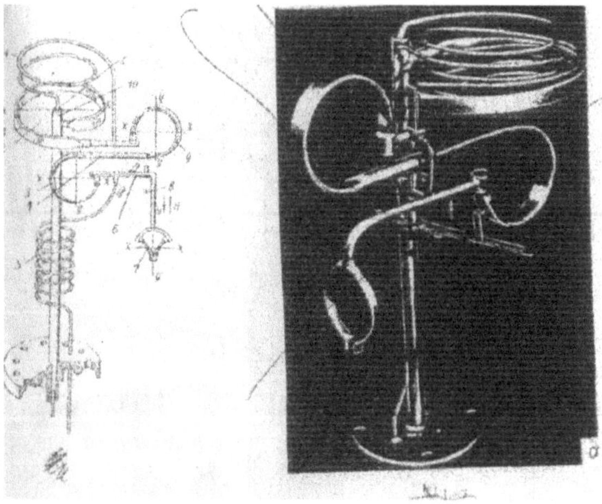

Fig. 9.6a Manipulator: (left) scheme, (right) general view.

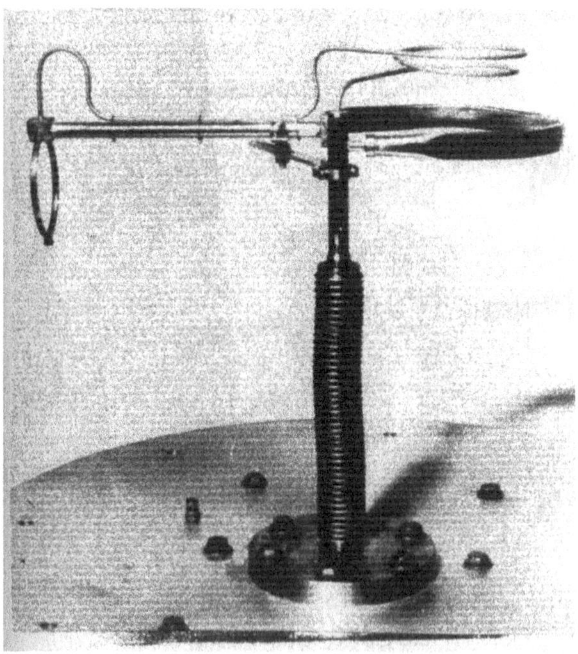

Fig. 9.6b The manipulator based on unclosed contour for usage in multifunctional devices of vacuum mechanics.

Fig. 9.7 Four position transfer manipulator of wafers.

Figure 9.6 shows an example of a drive of unclosed type which can be used for development of various vacuum mechanisms. This drive has three degrees of freedom.

The clamp 7 can turn around a vertical axis by driving element 2 and can be displaced in vertical direction by the elements 3 and 9. These elements are made of hermetically sealed tubes which allows us to extend the displacement range.

The sample capture is realized using clamp 7, which consists of a collector and two elastically deformed driving elements. Each motion is realized by the pressure control separately in each driving element through the tubes 10, 1 and 5. The tubes 1 and 5 contain spiral compensators. After angular transference the clamp position is fixed by supports 8.

A positioning manipulator for transfer of waters and other flat samples with diameter up to 200 mm was designed. The maximum angular displacement of this manipulator exceeds 90 degrees. The manipulator is based on a multi-turn drive which performs a near-circular motion (see Figure 9.7).

The manipulator is combined of four driving elements. Also there is a bellow pneumodrive for vertical displacement of the clamp (Figures 9.7 and 9.8).

The total angular displacement $\Delta_{\gamma\Sigma}$ of the free end of the drive is proportional to the number of coils and can be calculated from the following equation:

$$\Delta_{\gamma\Sigma} = n\gamma \frac{PR_0^2(D_6 + D_7\nu)}{Eah(D_3 + D_4\nu + D_5\nu^2)},$$

where γ is the central angle of the drive.

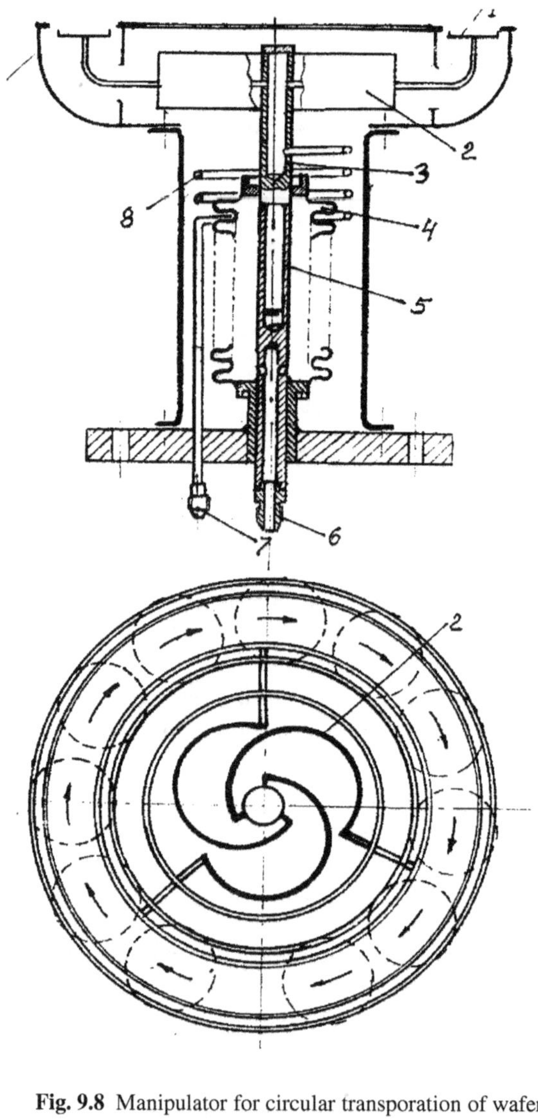

Fig. 9.8 Manipulator for circular transporation of wafers.

The turning drives shown in Figure 9.1b can be used in transporting systems of discrete action of multiposition vacuum equipment.

The scheme of a transporting device for sample displacement is shown in Figure 9.8. The drive of the device is a drive of discrete circular displacement 2 and consists of a bearing ring 1, guides 9 and bellows drive 4 of vertical transference. Bearing ring 1 moves in circular and in vertical directions and realizes the following operations: "taking – transferring – laying down". Gas supply of the bellows drive

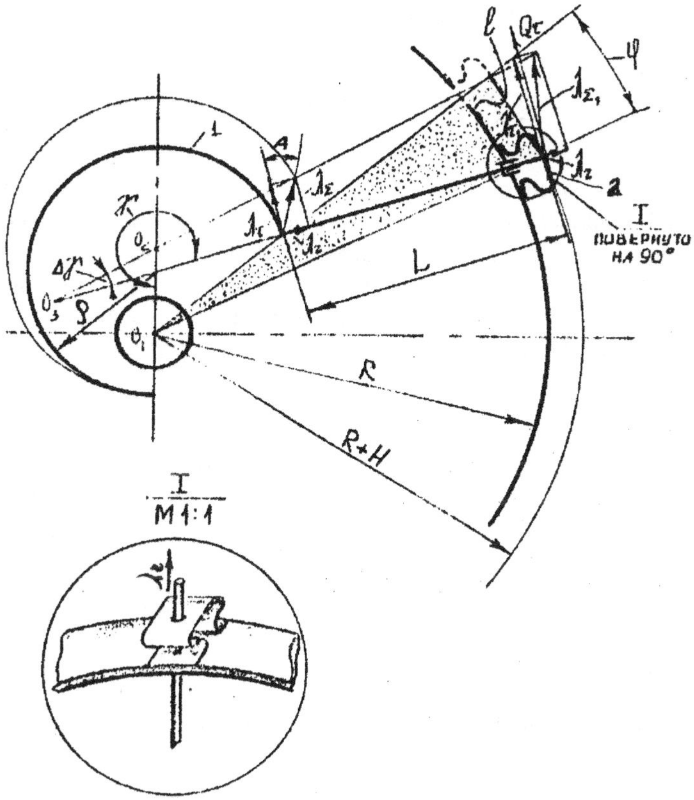

Fig. 9.9 Scheme for determination of the angular circular displacement.

is realized through the connecting pipe 7, and the tube with compensator 8 and the tube collector 3.

The scheme for determination of angular and circular displacement is shown in Figure 9.9. The circular errors, Q_1, Q_2 and Q_3, correspond to center of rotating device, initial center and momentary center of the curve of elastically deformed drive.

As follows from Figure 9.9 the angle of the ring rotation can be determined by

$$\varphi = 2 \arcsin \frac{l}{2(R+H)},$$

where R is the radius of outer ring of the drive; H is the height of the compensator. For engineering calculation it can be assumed that that

$$l = \lambda \tau_1,$$

where $\lambda\tau_1$ is the tangential component of the lever of length L. The value $\lambda\tau_1$ can be determined from:

$$\lambda\tau_1 = \frac{L\sin\Delta\gamma + \lambda_\Sigma \sin\Delta\gamma + \lambda_\Sigma \cos(A+\Delta\gamma)}{\lambda_\Sigma \cos(A+\Delta\gamma) + \lambda_r \sin\Delta\gamma}\lambda_\tau,$$

where $\Delta\gamma$ is the variation of the central angle γ of the driving tube; λ_τ, λ_r are the tangential and radial components of displacement which is formed by the free end of the tube element: λ_Σ is the total displacement of the free end of the tube element; A is the angle between the vector of the total displacement λ_Σ and the tangent vector of the central axis of the tube element. The values of the parameters λ_τ, λ_r and λ_Σ may be determined from the following equations:

$$\lambda_\tau = \frac{\Delta\gamma}{\gamma}R(\gamma - \sin\gamma),$$

$$\lambda_r = \frac{\Delta\gamma}{\gamma}P(1 - \cos\gamma),$$

$$\lambda_\Sigma = \sqrt{\gamma_\tau^2 + \lambda_r^2},$$

$$A = \arctan\frac{(1-\cos\gamma)}{(1-\sin\gamma)}.$$

Torque of the ring M has three components of loading force. These components coincide with the direction of tangential component of the torque:

$$M = 3Q\tau R.$$

The value Q_τ can be determined from Equation (9.2) taking into account the transmission ratio of the lever L. For the termination of the motion in radial direction λ_r, the spring compensators (pos. 2) are used. These compensators are fixed to the end of elastically deformed elements. The rigidity of these compensators in circle direction must be high enough for torque transmission.

The drives with a closed contour are the simplest and most useful from a technological point of view (see Figure 9.1). In the drives of these types there are four hermetic curved tubes of constant radius of curvature. The tubes are connected one to another and form a plane contour.

The scheme of the drive is shown in Figure 9.10. The resulting displacement in central point B along axis Y is determined by the total displacement deformation of the total perimeter of contour S. In the general case the resulting displacement can be determined by the integral:

$$\lambda_B = \int \frac{\phi(\varphi)}{\rho(\varphi)} M_1(\varphi)\,dS,$$

where $M_1(\varphi)$ is the bending moment in arbitrary cross-section of the tube as a function of unity force in the direction of specific displacement; $\psi(\varphi)$ is the angle of tor-

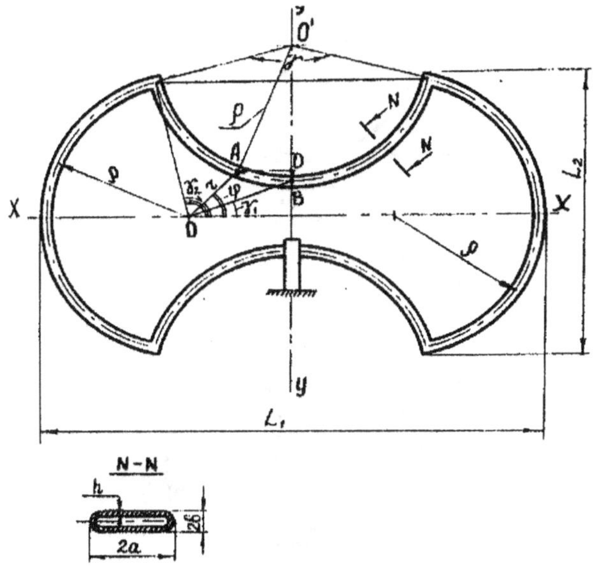

Fig. 9.10 The scheme for calculation of the drive of the closed type.

Table 9.2 The angle parameters (in degrees).

γ	90	120	150	210	240	270
γ_1	90	45	17.6	–13.4	–24.9	–35.3
γ_2	135	120	105	75	60	45

sion in the arbitrary cross-section; $\rho(\varphi)$ is the curve radius of contour element. The geometry parameters of elastic line of the contour depends on the angles γ, γ_1, γ_2:

$$\gamma_2 = \pi - \gamma/2.$$

The angle γ_1 is determined from the triangle OBC:

$$\gamma_1 = \arctan \frac{\sin\gamma/2 + \cos\gamma/2 - 1}{\sin\gamma/2 - \cos\gamma/2}$$

Table 9.2 shows several values of the angles. Figure 9.11 shows the contour form for two cases:

$$\gamma \leq 180^\circ \quad \text{and} \quad \gamma > 180^\circ.$$

The drive generate a force in the direction of the displacement which can be expressed in general form as

$$Q = \frac{EKI_a \int_S \frac{\phi(\varphi)}{\rho\varphi}[r(\gamma_1)\cos\gamma_1 - r\varphi\cos\varphi]dS}{1-\mu^2 \int_S [r(\gamma_1)\cos\gamma_1 - r(\varphi)\cos\varphi]^2 dS},$$

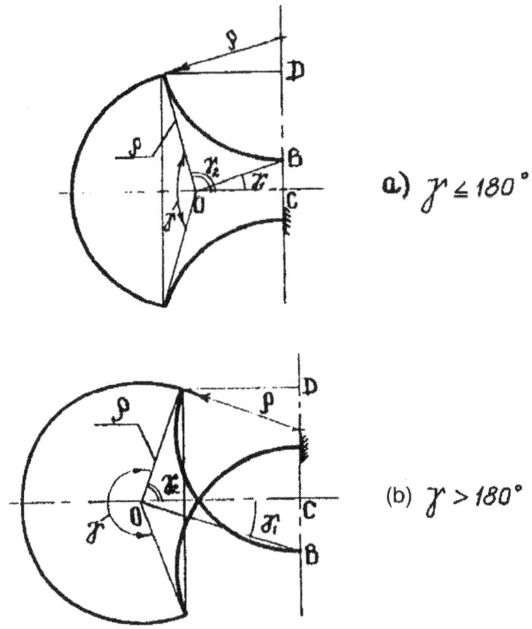

Fig. 9.11 Drives contours with different L: (a) $\gamma \leq 180°$; (b) $\gamma > 180°$.

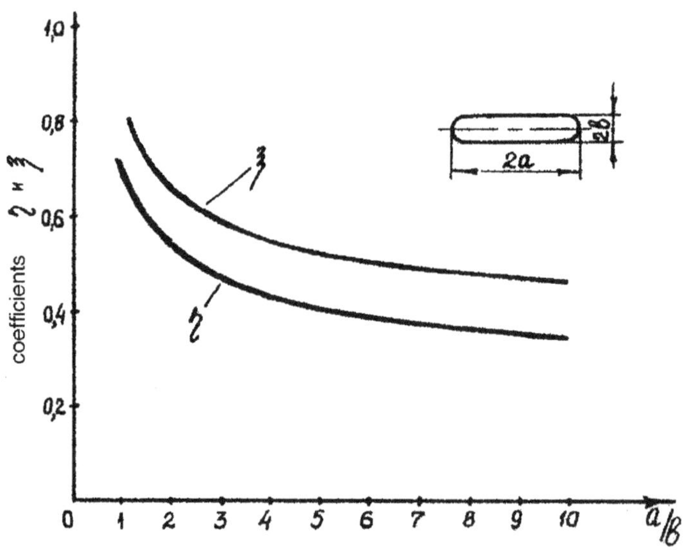

Fig. 9.12 The values of coefficients η, ξ for plane-oval cross-section.

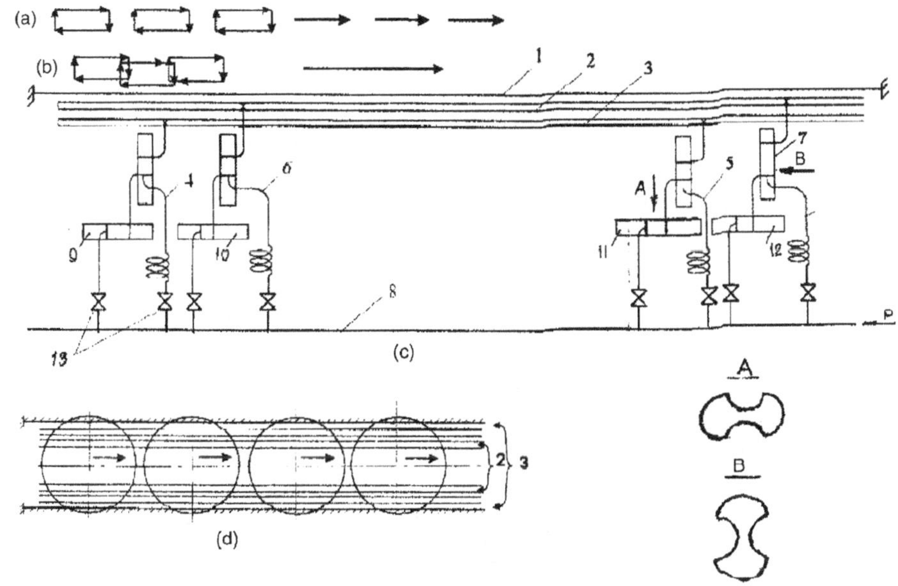

Fig. 9.13 Scheme of transporting system.

where r is a parameter shown in Figure 9.11; E is the modulus of elasticity of the drive elements material; μ is the Poisson coefficient; I_a is the moment of inertia of cross-section in respect to the large axis; K is the coefficient depending on the deformation of the cross-section element and curvature of contour of the central axis.

For the most commonly used plane-oval cross-section we have the following equation:

$$I_a = 4_{B^3h}(a/b - 1 + \pi/4),$$
$$K = 1 - \eta/(\xi + \chi^2).$$

where χ is the main parameter of the element $\chi = \rho h/a^2$; η, ξ are the coefficients which depend on the shape and the axis ratio of the normal section and the cross-section. Coefficients η, ξ for plane-oval cross-section are shown in Figure 9.12.

The drives of loop-controlled contour have high rigidity. These drives have enhanced dynamic parameters and a high damping factor of damped oscillation. They can be used successfully for linear transportation of samples. However, the direction of displacement trajectories depends on the geometry of the drive axis in space.

Figure 9.13 shows a variant of a transportation system which is designed on the basis of a loop-controlled drive contour. This drive allows intermittent (discrete) and streaming modes of operation. In the first case (Figure 9.13a) the drive realizes the operation mode "taking – transferring – laying down" and in the second case

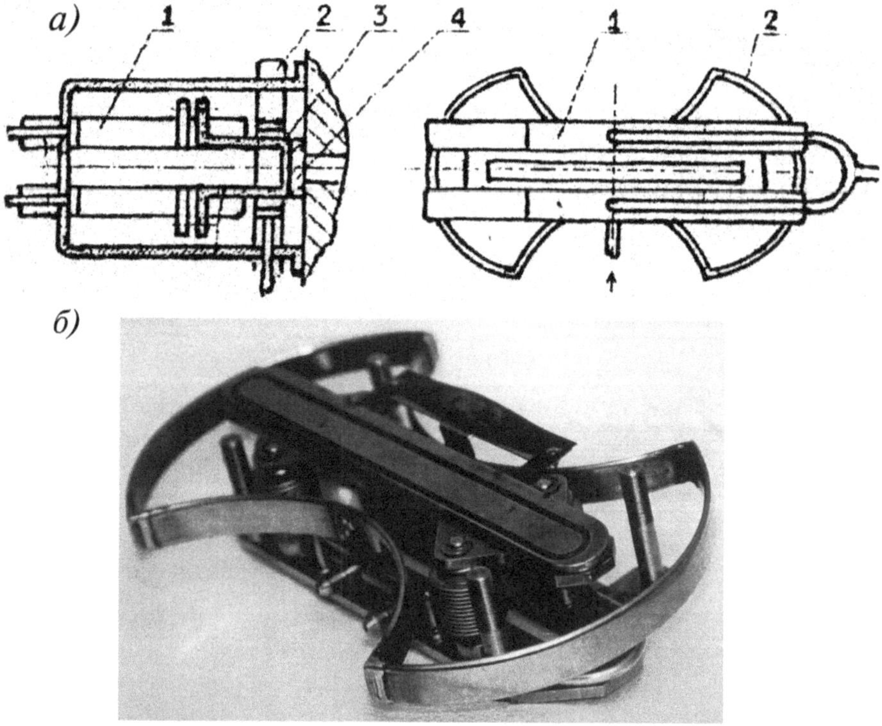

Fig. 9.14 Bodiless vacuum gate of slit type: (a) scheme of design, (b) general view.

(Figure 9.13b) the drive realizes the operation "taking – transferring". In the last case the displacement can be realized on required distance because of superposition of driving impulses. In the case of the intermittent mode of operation the drives 4, 5, 9, 11 are used; in the case of the streaming mode of operation the drives 4 to 7 and 9 to 12 are used. The system consists of two pairs of mobile guides 2 and 3 placed along a static guide 1.

A vertical displacement of movable guides 2 and 3 is realized using the drives 4, 5, 6, 7; the horizontal transference is realized by the drives 9, 10, 11, 12. Compressed gas or liquid are supplied from the common collector 8. Sequences of switching of electromagnetic valves 13 is controlled by a program according to the required regime of the operation.

The speed of transference can reach 100 mm/s in a streaming mode of operation. By using liquid the 2 or 3 coordinate precise simple drive with loop control can be developed. The drives were used for aperture diaphragms in electron beam installations and other electronic technological equipment. Figure 9.14 shows the scheme of a loop-controlled contour of the drive of bodiless vacuum gate. This type of vacuum gate can be used for isolation of a vacuum system sealing under small pressure difference.

In a vacuum gate of this type, two drives 1 create the sealing load through the clamping cramp 3. The drive 2 through the flexible element rises sealing strap 4 in vertical direction and opens the slit hole. The considered drives can be manufactured from thin-wall tubes with wall thickness 0.15–0.2 mm. These thin-wall tubes are made by a solid-drawn method and then the required normal cross-section and curve radius of central axis are obtained by deformation. Steel used for manufacturing of the driving tube elements have good elastically durability, vacuum technological and processing characteristics. The following steels can be recommended for manufacturing of thin-wall driving tubes: 36NiCrTiAl, 2Cr13, 08Cr18Ni10Ti.

Chapter 10
Flow of Microparticles Originating from Mechanisms in Vacuum

10.1 Theory of a Flow of Microparticles Originating from Mechanisms in Vacuum

The theoretical foundation of the formation and distribution of the flow of small dispersible particles in vacuum [1] creates some insight in different sorts of behaviour of powder-like micro particles at normal atmospheric pressure and at different pressure ranges: from normal atmospheric pressure to high vacuum. The behaviour of powder-like micro particles as a result of the action of residual gas molecules is also discussed. The basic positions of the theory can be formulated as the following:

(1) The microparticle flows can be divided into primary and secondary ones. The secondary flows are formed from monolithic materials and from solid films during mechanism operation in vacuum.

Their interaction with residual gas environment depends on random collisions with gas molecules. The secondary flows are formed from primary microparticles which are detached from surfaces inside vacuum chambers and which are transferred together with the pumped gas or together with an inlet gas flow in a process of aerodynamic interaction.

(2) The process of newly formed debris as a result of primary flow action can be presented in three stages (Figure 10.1): generation of microparticles, transport of microparticles from their source to the surface of the product. The generation of small dispersible particles is a random process which is described by the corresponding laws of distribution of microparticles number, size d_p distribution, distribution of direction and velocity V_p of microparticles detachment, electric charge and magnetic moment distribution.

A microparticle in the process of transference in high vacuum is subjected to resistance force F_r of environment. This force is directed opposite to the direction of the movement of microparticles; the resistance force varies in time and it also depends on the pressure of the environment. The dispersion of microparticles

E.A. Deulin et al., *Mechanics and Physics of Precise Vacuum Mechanisms*,
Fluid Mechanics and Its Applications 91, DOI 10.1007/978-90-481-2520-3_10,

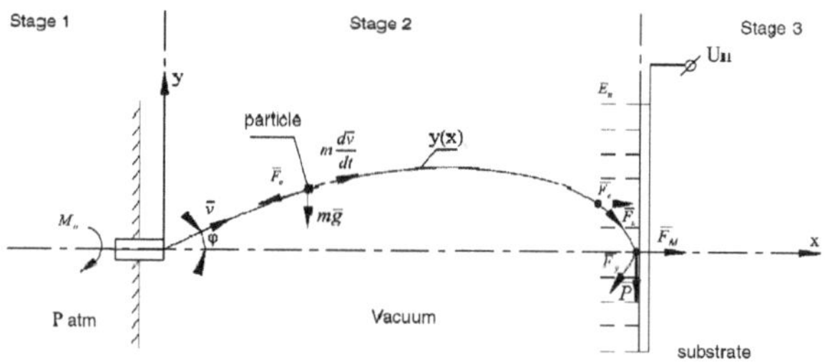

Fig. 10.1 Scheme of primary flows of small dispersible particles forming.

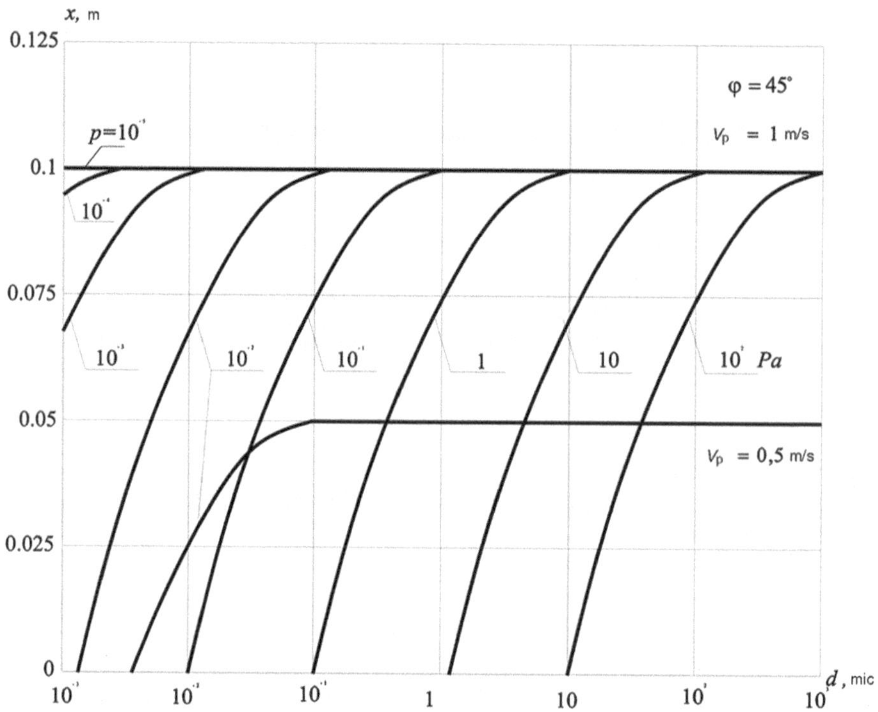

Fig. 10.2 Graph of flight length x of small dispersible particles as a function of their diameter d_p, velocity V_p, angle φ of detaching, also as a function of pressure p in the vacuum chamber.

on the surface of the product depends on the electric charge or on the magnetic moment of small dispersible particles (F_m), product potential U_w and E_w, inertia force F_i, P, adhesive force (molecular interaction force F_m) as well as on elasticity force F_e, which appears as a result of a collision of microparticles with the substrate. Figure 10.2 shows the example of calculation of the flight length x of the microparticles generated by vacuum mechanism at different velocities V_p of detachment and at different pressures p in a vacuum chamber.

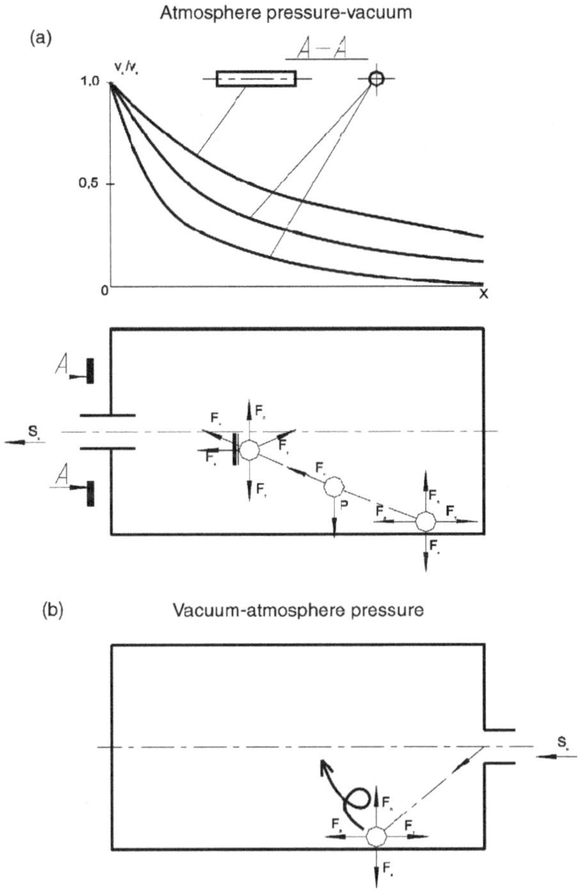

Fig. 10.3 Scheme of vacuum chamber and corresponding scheme of the graph V_x / V_0 of velocities distribution of a gas which is pumped from atmosphere pressure in case of different forms of pumping tubes. (a) Scheme shows microparticles which are fixed on the surface of chamber which are transferred inside the chambers volume and which are impacted with the surface of the product. (b) Scheme shows microparticle fixed on the surface which is loaded with the forces which acts on the microparticle in process of gas lap joint into the vacuum chamber.

(3) The process of newly formed debris as a result of secondary flows of small dispersible particles action (as in the previous case) can be presented in three phases (Figure 10.3a): they detach from the surface, they move together with the gas and then land on the substrate. The intensity of these phenomena depends on the adhesive interaction F_a of micro particles with the surface of a solid body, their weight P, frontal resistance force F_h of the pumped gas, resistance force of friction of the surface F_f, carrying capacity force F_c, inertia force F_i, elasticity force F_e, in the case of a microparticle impact with the surface of a solid. It depends also on the gradient of the gas flow movement decreasing during pumping process from at-

mosphere pressure V_x/V_0, the form and sizes of tubes, the effective pumping speed and other factors.

The gas input to vent a vacuum chamber (Figure 10.3b) has a large influence on the process of small dispersible particles detachment from the walls of the vacuum chamber and from the surfaces of mechanical elements which are situated inside vacuum chamber. This is because the thickness of a border film on a surface is close to zero and microparticles are not "protected" by this film in high vacuum.

The sizes of detached small dispersible particles are in the range which depends on the effective pumping speed S_0 (Figure 10.4a) and the gas input at speed S_i.

This range increases when S_0 or S_i increase. This is a reason for the phenomenon which is illustrated in Figure 10.4b: the number of detached microparticles increases and the lowest possible sizes depend on the difference between the adhesion force of the microparticle and the frontal resistance force of gas; the highest limit of the size depends on the increase of microparticle weight.

The condition of a microparticle movement in the volume of a vacuum chamber and the conditions for microparticle movement in a vacuum system in a pumping process is under the condition that the speed of the gas flow is higher than the static speed of the precipitation of small dispersible particles.

The part of the flow of microparticles which are detached from vacuum surfaces can be precipitated on the substrate surface or can be bounced off from this surface after collision with this surface due to elasticity.

The landing for the smallest microparticles is "simpler" than for bigger microparticles because the probability of the adhesive force of very small microparticles is higher than the elasticity force.

The graph of the number of precipitated microparticles depends on an effective pumping speed. This graph has the turning point as it is shown in Figure 10.4.

So, the microparticles of silicon carbide which are fixed on a point of a vacuum chamber surface (in case of tube diameter 0.04 m) and the microparticles fixed in point coordinates $y = 0$ and $x = 0.2$ m; the detached microparticles diameter range is from 20 to 30 mic (in case $S_0 = 1$ l/s) and the detached microparticles diameter range is from 0.001 to 120 mic (in the case of $S_0 = 10$ l/s) as shown in Figure 10.4a.

In the case $S_0 = 1$ l/s, the gas flow detaches the microparticles with diameters below 10 mic and for $S_0 = 10$ l/s the coarse-grained (bigger) microparticles with diameters $d_p < 30$ mic can be detached.

After the microparticles collide with an obstacle, the microparticles stay on the obstacle surface and, in the case $S_0 = 1$ l/s, the diameter of precipitated microparticles is below 0.1 mic and for $S_0 = 10$ l/s the microparticle diameter is below 0.02 mic.

For the pumping speed $S_0 = 1$ l/s, the number of microparticles which can be detached from the surface at the distance $x = 0.2$ m from the tube hole, is about 18% of the number of small particles with diameter in the range from 0.001 to 120 mic. When the pumping speed is $S_0 = 10$ l/s, the number of microparticles which can be detached from the surface is about 47% of small particles (Figure 10.4b). When the pumping speed is $S_0 = 1$ l/s, the number of microparticles which can be detached from the surface is about 18% of all particles detached from the surface.

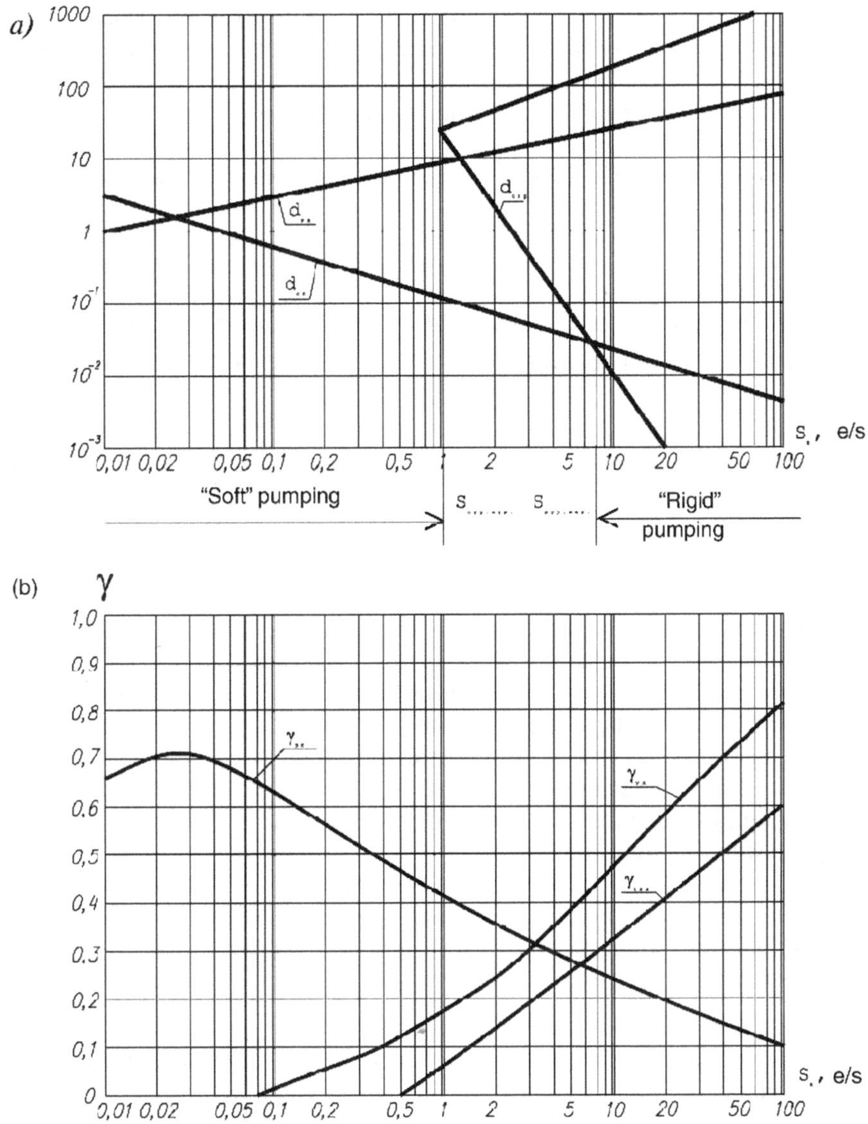

Fig. 10.4 Graphs of small dispersible particles sizes (a) and of small dispersible particles portion (b) which are detached from the vacuum chamber surface (d_{det} and γ_{det}), which are removed from the vacuum chamber volume (d_{rem} and γ_{rem}) wich are precipitated on the substrate (d_{pre}, γ_{pre}) as a function of the effective pumping speed S_0.

When the pumping speed is $S_0 = 10$ l/s, the number of microparticles which can be detached from the surface is about 32% of the small particles.

For a pumping speed of $S_0 = 1$ l/s, the number of microparticles which can adhere to the obstacle is about 40% of all detached microparticles and for the pumping speed $S_0 = 10$ l/s the number of these microparticles is only 24% of the total number of small particles. The maximum number of microparticles which can adhere to the obstacle is in the case when $d_{pre} = d_{rem}$ and it corresponds to the pumping speed of a vacuum system at value S_0. In this case, the number of small particles is at its maximum.

10.2 The Design of the Equipment Which Generates a Minimal Number of Microparticles by the Mechanisms in Vacuum

Microparticles generation by mechanisms operating in vacuum can be decreased by the following measures: selection of antifriction and wear-resistant protected materials; use of special technological methods of machine details fabrication; design of the mechanisms which generate a minimum flow of small particles [2].

Hard lubricants (SLC-solid lubricant coating) manufactured by the Russian IC R&D company, type "Dimolit-4", wear-resistant materials (nitrides, carbides, silicides of metals) and anti-friction materials (teflon, polymide and others) are used in electronic machinery: in rods, ways, ball bearing rolling and sliding carriers.

To ensure a minimal generation of powder-like particles, the source analysis was done (Table 10.1). It has been shown that the most promising coating is molybden disulphide deposited by ion-plasma sputtering in vacuum.

The flow of small particles from a roller-ball bearing is covered with this coating is 200–800 times smaller than the flow from the same roller ball bearings which are not covered with solid lubricants.

After a longer work period the flow of microparticles from bearings without coatings increases. However, the flow of microparticles decreases from bearings covered with solid lubricants (until the cover film will wear out). To decrease the flow of generated microparticles, non-traditional methods of machining are used:

- plasma-chemical modification of rubbers with teflon,
- vacuum ion-plasma sputtering (IPS) and solid lubricant precipitation method of roller ball bearings and of sliding bearings in vacuum,
- magneto-impulse processing of ferromagnetic dusts to achieve magnetic liquids,
- and other methods.

The best solution to the problem can be to use mechanisms in vacuum without friction couples. These mechanisms are based on controlled elastic deformation. Nevertheless, these mechanisms have worse characteristics, rigidity, precision, dimensions than traditional drive mechanisms.

There is a great number of standard vacuum motion feedthroughs: bellow sealed feedthroughs, harmonic drive feedthroughs, original designs based on eutectic alloy and on magnetic compositions. Only a drive supplied with a magnetic liquid seal has the smallest flow of generated small dispersible microparticles. The main

Table 10.1 Summarized flow N_p (numbers/sec) and middle size d_p (mic) of small dispersible particles generated by different units and mechanisms in vacuum.

Unit or mechanisms	N_p numbers/s	d_p mic
Rotation feedthrough with magnetic fluid seal ($n = 100$ rev/s)	10^{-7}	1.5
Ball bearing with IPS MoS_2	10^{-4}	1.6
Rotation feedthrough with magnetic fluid seal ($n = 100$ rev/s)	$2 * 10^{-4}$	0.8
Elastomer rotation feedthrough with magnetic fluid seal ($n = 100$ rev/s)	10^{-3}	1.0
Conveyor belt	$7 * 10^{-2}$	0.75
Drum with substrates	1.2	0.5
Sliding process of substrates in installation (in Russian Vesuvi-13)	3.1	0.5
Rolling ball bearing without lubricant	10	0.8
Sliding bearing with hard lubricant type "DIMOLIT-4"	10^4	3.0
Substrates holder as merry-go-round	$1.26 * 10^5$	0.8
Jack ladder	$1.7 * 10^6$	1.5

recommendations in design and equipment are to decrease defects of the worked substrate surface. This recommendation can be formulated as:

> When substrates are present in a vacuum chamber (Figure 10.5a) it is necessary to realize "soft" pumping or gas input with pumping speed $S_0 < S_{0,\min}$ [2] (see Figure 10.4a). It is better for this process to ensure that the chamber is connected with the tubes, connected as far as possible from the bottom. Tubes entering into a chamber should be at the distance 1.5–2 diameters from the bottom. It is necessary to cool the walls of a chamber to a temperature of 8–10°C to increase adhesive forces between microparticles and the surfaces, also as it is necessary to put more light gases in, for example, nitrogen or helium.

If the walls of the chamber, or the bottom, or mechanic elements inside vacuum chambers are contaminated with the wear products/debris and if the surfaces are covered with condensed liquid it is necessary to clean the chamber using a "rigid" pumping process with pumping speed $S_0 > S_{0,\max}$ (Figure 10.4a).

In this case, it is better to use input tubes designed in the form of slit with cross-section sizes ratio 1:5 to 1:10 and to mount them as close to the bottom as possible. Also, it is necessary to heat the walls of the chamber to a temperature of 8–10°C to decrease the adhesive forces between microparticles and the surfaces, also as it is necessary to add more dry gases with larger molar mass than air, for example, argon or krypton.

To avoid microparticles from directly entering into the substrate, the mechanism and other sources of small dispersed particles used in vacuum must be specially positioned. Screens, labyrinths, magnetic and electrostatic catchers can be used for this task. The substrates best position is vertical (see Figure 10.5) and the best place for their transportation is the upper part of the vacuum chamber. The best way of transport is stepped transport or transport by mechanical arms. Before the processing starts, the microparticles precipitated on substrates can be removed from the sub-

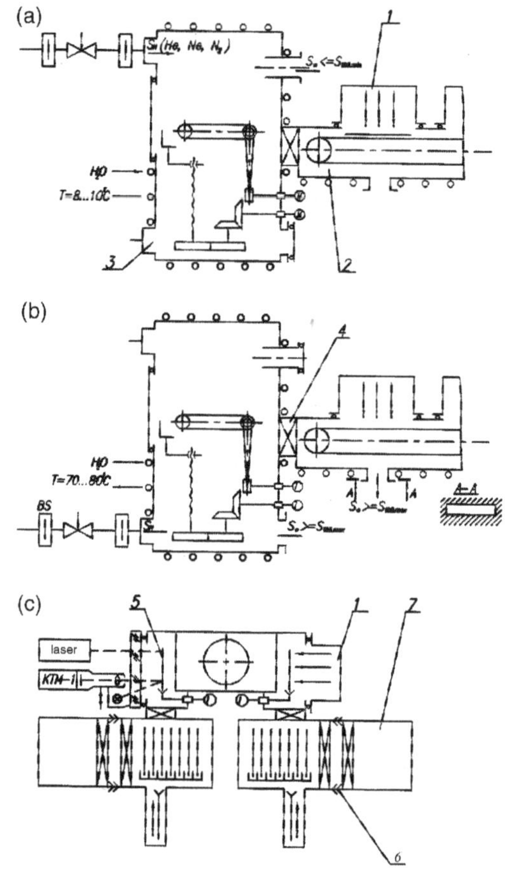

Fig. 10.5 Recomendations in design of vacuum technological equipment according criterion of minimum introduced unsoundness (defects): 1. work source, 2. vacuum transporting system, 3. sluice loading device, 4. vacuum gate, 5. plate, 6. vacuum container, 7. SMIF-interface.

strates using mechanical, chemical, aerodynamic, electric, magnetic, or laser action (Figure 10.5b).

Analysis and control of flows of microparticles landed on the substrate as well as analysis and control of the efficiency of removing microparticles, can be done using optical devices. For example, it can be done using a television camera KTM-1 supplied with matrix photodetector.

One of the main principles of design of multichamber cluster-type vacuum installations [3] is to choose the correct working principle, and a principle of selection of inter-operational transport which ensures a high product quality (low defect level). The cluster-type vacuum installation is used for ultralarge integral schemes manufacturing. This principle can be realized in two ways:

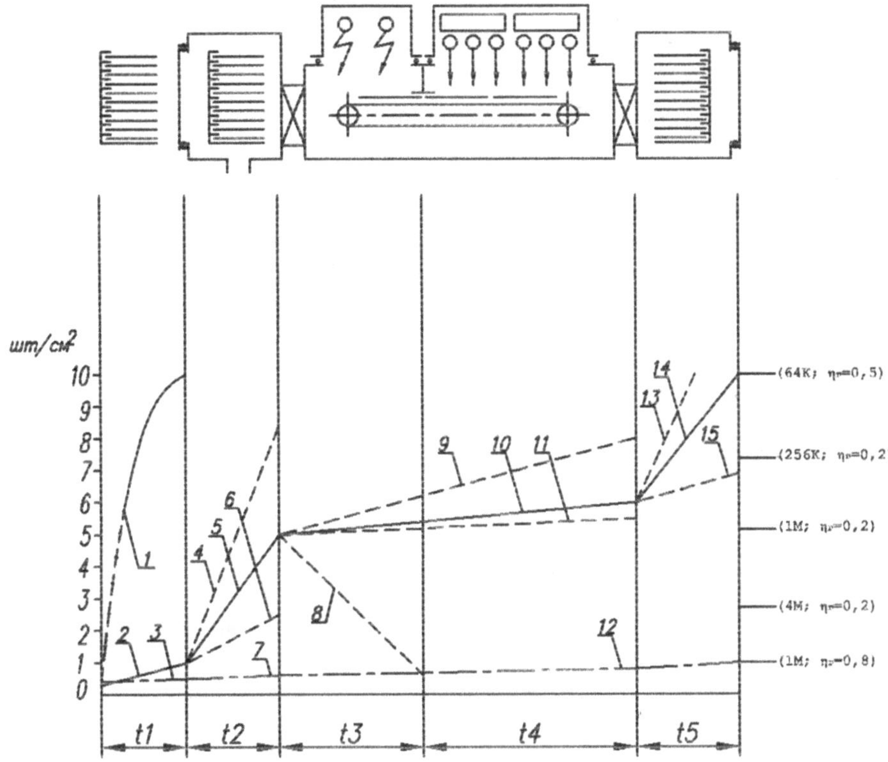

Fig. 10.6 Scheme of storaging process in criterion of unsoundness level control of processed product: 1. class 100; 2. class 10; 3. class 1; 4. $\gamma_{oc} = 0.2$; 5. $\gamma_{pre} = 1.17$; 6. $\gamma_{pre} = 0.1$; 7. $\gamma_{pre} = 0$ (SMIF); 8. N_p; 9. $N_p = 1 \times 10^{-1}$ numbers/ (sm^2·s); 10. $N_p = 5 \times 10^{-2}$ numbers/(sm^2·s); 11. $N_p = 1 \times 10^{-2}$ numbers/(sm^2·s); 12. $N_p = 1 \times 10^{-2}$ numbers/(sm^2·s); 13. $\gamma_{pre} = 0.2$; 14. $\gamma_{pre} = 0.17$; 15. $\gamma_{pre} = 0.1$; 16. = 0 (SMIF).

- The first way is the traditional way of mechanical engineering and it consists in one by one positioning of vacuum technological modules and transporting systems.
- The second way is non-traditional. This way corresponds to astronautic "expeditions of attendance" and corresponds to object entry into a vacuum container which is supplied with standard mechanical interface SMIF-interface (Figure 10.5b).

On the one hand, the realization of these methods allows us to decrease the introduced imperfection. And on the other hand, the SMIF-interface helps to radically decrease the introduced inperfection. This method helps to exclude the secondary flow of small dispersed particles to enter into the substrate, the pressures in the container and in vacuum chamber are equal and this method is highly reliable (reliability in functioning process). This method corresponds to minimal stops underway.

Apart from the afore-mentioned methods, the container can be used for interoperational storage of unlimited capacity. It can also be used as an interoperational transporting device.

Considering these conditions, the requirements for cleanness of manufacturing volume (“clean room volume”) can be transformed into container volume.

Regulation of the storing process of introduced imperfections (Figure 10.6) can be done using a selection of the cleanness range, transporting method and loading the substrate in selecting the installation method, vacuum transporting system selection, substrate surfaces cleaning of microparticles devices. All these methods must correspond to the process of suitable microschemes manufacturing.

References

1. Panfilov Y.V., The theoretical basis of small dispersed particles forming and distribution in vacuum, in *BMSTU Proceedings*, Mashinostroenie, Moscow, Vol. 2, 1993, pp. 87–94 [in Russian].
2. Panfilov Y.V., Design of vacuum technological equipment for ultralarge integral schemes manufacturing according criteria of minimal introduced unsoundness, *Electronic Technic., Serial 3*, Microelectronics, Moscow, 1990, Vol. 1 (135), pp. 51–59 [in Russian].
3. Malgin S.N., Panfilov Y.V., Cluster equipment in electronics, *Electronic Technology Reviews, Serial 7*, 1994, Vol. 1 (1701), 120 pp. [in Russian].

FSC
www.fsc.org
MIX
Papier | Fördert
gute Waldnutzung
FSC® C083411

Kolibri 360 GmbH
Ferdinand-Jühlke-Straße 7
99095 Erfurt, Deutschland
produktsicherheit@kolibri360.de